A Treatise on the Tactical Use of the Three Arms: Infantry, Artillery, and Cavalry

by Francis J. Lippitt

I0493070

TO THE MILITARY PUBLIC.

The Author would feel obliged for any facts or suggestions which might enable him to render a future edition of this work more valuable.

PROVIDENCE, R.I., July, 1865.

TABLE OF CONTENTS.

Tactical Use of Infantry

I. ITS ATTACK, GENERALLY II. FORMATIONS FOR ATTACK III. THE ATTACK, HOW MADE IV. BAYONET CHARGES V. DEFENCE AGAINST INFANTRY VI. DEFENCE AGAINST ARTILLERY VII. DEFENCE AGAINST CAVALRY VIII. SQUARES IX. SKIRMISHERS A. THEIR USE B. HOW POSTED C. HOW HANDLED D. RULES FOR INDIVIDUAL SKIRMISHERS

Tactical Use of Artillery

I. HOW POSTED WITH RESPECT TO THE GROUND II. HOW POSTED WITH RESPECT TO OUR OWN TROOPS III. HOW POSTED WITH RESPECT TO THE ENEMY IV. POSTING OF BATTERIES AND OF PIECES AS BETWEEN THEMSELVES V. HOW USED A. GENERALLY B. IN OFFENSIVE COMBAT C. IN DEFENSIVE COMBAT D. AGAINST INFANTRY E. AGAINST CAVALRY F. AGAINST ARTILLERY VI. ITS FIRE VII. ITS SUPPORTS

Tactical Use of Cavalry

I. ITS FORMATIONS II. ITS STRONG AND ITS WEAK POINTS III. HOW POSTED IV. ITS SUPPORTS V. HOW USED VI. HOW IT FIGHTS VII. ITS CHARGE VIII. ITS ATTACK ON INFANTRY A. GENERALLY B. ON SQUARES IX. GENERAL REMARKS

TACTICAL USE OF THE THREE ARMS.

Every complete military force consists of three arms,--INFANTRY, ARTILLERY, and CAVALRY.

In battle, these three arms are united; and, other things being equal, that commander will prove victorious who is best acquainted with their combined use in the field.

In order thoroughly to understand the proper use of the three arms combined, we must obviously begin by learning the proper use of each of them separately.

Hence the importance of the subject of the present treatise. In discussing it, we shall commence with the

TACTICAL USE OF INFANTRY.

The subject will be considered under the following heads:--

I.--ITS ATTACK, GENERALLY. II.--FORMATIONS FOR ATTACK. III.--THE ATTACK, HOW MADE. IV.--BAYONET CHARGES. V.--DEFENCE AGAINST INFANTRY. VI.-- DEFENCE AGAINST ARTILLERY. VII.--DEFENCE AGAINST CAVALRY. VIII.-- SQUARES. IX.--SKIRMISHERS.

I.--Its Attack, generally.

Infantry attacks with its fire, or with the bayonet. Which of these is the more effective?

1. The object of an attack is to destroy or capture the hostile force, or, at least, to drive it from the field.

Capturing the enemy, or driving him from the field, cannot usually be effected by merely firing upon him.

True, a mere fire at a distance may finally destroy him. But an insuperable objection to this mode of attack is, that while we are killing or disabling his men, he is killing or disabling as many of our own.

2. If we fire from behind cover, our loss may be comparatively small. But, in that case, the enemy will never remain for any length of time exposed to our fire. He will either attack and rout us from our cover, or retire. And even if he

did neither, his actual and complete destruction, capture, or rout, would still require an attack with the bayonet.

3. It follows that the proper mode of attack by infantry on infantry is with the bayonet.

The Russian Suwarrow's victories and reputation were won chiefly by his fierce bayonet attacks, which often effected great results, in spite of his ignorance of the art of war.

4. But there are exceptional cases where infantry may properly use only its fire; as--

(1.) When acting as a support to artillery, it should rarely, if ever, leave its position to use the bayonet; thereby endangering the safety of the guns which it is its first duty to guard. Its function, in this case, being purely defensive, it should act by its fire alone.

(2.) Against a line of skirmishers deployed, a well-directed fire will usually be sufficiently effective.

(3.) In mountain warfare, its only practicable mode of attack will sometimes be by its fire.

5. When both sides are equally exposed, the actual attack with the bayonet should not be preceded by a distant musketry fire; for, as in that case, our loss will generally be equal to the enemy's, this fire will give us no superiority in the charge, and the loss we have sustained will be therefore entirely thrown away.

6. Nevertheless, our actual attack should be prepared, when possible, by the infliction of such a loss on the enemy as will make him inferior to us at the decisive moment. In war, the object is not to test the comparative courage of the combatants, but to beat the enemy. We must never, therefore, when it

can be avoided, fight him on equal terms; and so, never close with him without such a superiority in numbers, position, or spirit, as will make the chances decidedly in our favor. If, without exposing ourselves to much loss, we can inflict a considerable loss upon him, we shall render him inferior to us, both by the number of his men we have disabled, and by the demoralization thereby caused in his ranks.

7. This preparatory loss can be most effectually inflicted by the fire of artillery; as, from its great superiority of range, it can suffer but little, meanwhile, from the enemy's infantry fire. Our attacking infantry are thus enabled to keep out of the range of the fire of the infantry they are to attack, till the moment of advancing to close.

8. When we have no artillery disposable for the purpose, the preparatory effect may be produced by a well-sustained fire of infantry, provided it can find a sheltered position to deliver it from; or, by the fire of a heavy line of skirmishers.

9. If we can make the infantry we wish to attack engage in a prolonged fire, this will exhaust them, and thus render them inferior to us in strength and in spirit, even if we inflict on them but little loss. But as our attacking infantry should, in the mean time, be kept fresh, the preparatory fire, in such case, should not devolve on the troops that are to close with the enemy.

10. One cause of the indecisiveness of the results obtained in many of the battles of the late war, as compared with the great loss of life on both sides, has been, that the opposing battalions were too often kept firing at each other at a distance, both sustaining nearly equal loss, until the ranks were so weakened as to disable either party from making a vigorous and decisive charge. Or else, charges were made on the enemy's battalions before they had been shattered by artillery; so that the attacking troops were easily repulsed, sometimes with great slaughter.

II.--Formations for Attack.

1. Infantry may advance to attack in either of three ways: in column; in line, marching by the front; and by the flank; that is, in line, but faced to a flank.

2. Of these three formations, the last is undoubtedly the worst possible; for-

-

(1.) On arriving at the enemy, the troops are not concentrated at the point where the struggle is to be. As they must come up successively, they will be crushed in detail by superior numbers.

(2.) Advancing in such a formation, they would be exposed to a destructive raking fire from the enemy's guns; especially since the adoption of the new flank march by fours, which gives to rifled artillery a tolerable mark.

3. The question is, then, between an attack in column and an attack in line. Which is the better of the two?

The decisive effect of infantry is produced by a rush on the enemy with the bayonet. The chief elements of success in this attack at close quarters are, the physical momentum of the charge, and the powerful moral effect caused by the swift approach of a compact and orderly hostile mass. A charge in line does not admit of both these elements. The advance of a line of one or more battalions, to be united and orderly, cannot be rapid, and thus has no impetus. Such a line, advancing swiftly, especially over uneven ground, would soon become so broken and disunited as to destroy, in a great measure, the effect, both moral and physical, of its charge, and, at the same time, to deprive the attacking troops of that confidence which is inspired by the consciousness of moving together in one compact, formidable mass, in which every soldier feels himself fortified by the support of his comrades.

4. On the other hand, a column can move rapidly without losing its compactness and order.

In attacking the enemy's line, a close column concentrates successively, but rapidly, a force superior to the enemy at the decisive point, and can hardly fail to pierce the line attacked, if it arrive with its momentum unchecked.

In a close column, there is a real force created by the pressure of the mass behind on the leading subdivision, pushing it on the enemy, and preventing it from drawing back or stopping; thus imparting to it somewhat of the actual physical momentum of a mechanical engine.

A close column shelters raw troops, and carries them irresistibly along with it.

A close column, in case of need, can rapidly extend its front by deploying.

It can promptly make itself impenetrable to cavalry.

Finally, in a column, the officers being seen by the men, the benefit of their example is not lost.

The close column would, therefore, seem to be the best formation for attack.

5. Movements in line requiring that high degree of perfection in drill which can rarely be attained by any but regular troops, they were accordingly abandoned by the raw and undisciplined masses of French soldiers that so successfully defended the French Republic from invasion against the veteran armies of Europe; some of which were led by generals who had served under Frederick the Great. Conscious of their military inferiority to the enemy, they instinctively clustered together in close and heavy columns; then rushed down on the enemy's line with the force of an avalanche, often carrying every thing before them. Thus was inaugurated that system of attack in deep and solid columns, which was afterwards so successfully used by Napoleon.

6. Close columns have two defects. One is, that they are oppressive and exhausting to the men, especially in hot weather.

But this is not a very serious objection; for they are, or should be, formed only when about to be used, and then their work is generally soon over.

7. The other defect, however, is of so grave a nature as, in the opinion of some, to more than outweigh their advantages; and this is, the terribly destructive effect upon them of the enemy's artillery fire, or of that of his sharpshooters; for the solid mass is an easy target, into which every shot is sure to penetrate. Many of the missiles which would fly over an advancing line, are sure to fall, somewhere or other, in a deep column.

This destructive effect was strikingly illustrated in Macdonald's charge on the Allied centre at Wagram. The eleven thousand men (some accounts say fifteen thousand) composing that famous column, advanced under the fire of one hundred and eighty hostile guns. After being driven back twice, they succeeded, in a third attack, in breaking the enemy's centre. But of the entire column, only eleven hundred men, it is said, were left standing.

8. The recent improvements in fire-arms must render the fire on a close column of infantry, both by artillery and sharpshooters, still more destructive than it was before. But this sacrifice of life can be prevented, to a great extent, by using the columns at a proper time and in a proper manner. They should, like storming parties (which they really are), never be launched against the enemy's line till the fire by which they would suffer has been quite or nearly silenced by our batteries. Sometimes this may be impracticable; but this precaution has often been neglected when it was perfectly feasible, thus causing a great and useless slaughter.

9. But destructive as may be artillery fire on close columns, on troops advancing in line grape and canister begin to be equally so on their arriving within four hundred yards of the enemy's batteries; and are certainly quite as destructive, and more so, at the distance of two hundred yards. So that, within this distance, at least, the superiority of lines over columns ceases; and, probably, much sooner.

10. The desideratum is to preserve the advantages of the column, while saving the attacking troops from the almost total destruction which would now seem to threaten them, when marching in such a formation, from the new rifled artillery, which is said to fire with accuracy at two thousand yards, and from the new infantry rifles, said to be reliable, in the hands of sharpshooters, at five hundred yards.

11. Perhaps this object might be attained by the advance of the attacking troops in line, but in loose order, and at double quick, to about two hundred paces from the enemy, a halt, a prompt alignment on the colors, a rapid ployment into close column doubled on the centre, followed by a swift and resolute charge with the bayonet.

This method, while giving the rapid clearing of the intervening ground, to within two hundred paces of the enemy, and afterwards the impetus, and other advantages of the column, would, at the same time, afford that comparative immunity from a destructive fire which is the chief advantage of an advance in line.

To guard against the danger, in the use of this method, of the troops stopping to fire, instead of ploying into a column of attack, they should commence their advance with pieces unloaded. Their boxes might even be previously emptied of their ammunition. Why should not a battle, as well as an assault on a fortress, have its "forlorn hope?"

12. This mode of attack would be open, it is true, to two objections:--

First. It would require for its successful execution under fire great coolness, and much previous instruction in the manoeuvre, to enable the troops to perform it promptly and accurately.

Secondly. In presence of a bold and active enemy, it would expose the attacking troops to the danger of being charged and routed while manoeuvring.

13. In the late War of the Rebellion, in lieu of close columns, attacks have been sometimes made in several lines, following each other at distances of three hundred paces or more. Although these attacks have sometimes succeeded, they are objectionable in principle; for each line is in danger of being repulsed successively, before the arrival of the one in its rear; and there is wanting that great superiority of force at the decisive point which is the most important element of success in a battle.

Such formations are essentially defensive in their nature, and not suitable for attack. A line in position, against which the enemy is advancing, is strong in its fire, which will usually preserve it from absolute defeat till a second line, posted at one hundred and fifty, or even three hundred paces in its rear, has had time to come up in support. But even these distances Napoleon's experience appears to have taught him to be much too great; for in his last battle, at Waterloo, he posted his second line, both infantry and cavalry, at only sixty paces behind the first; thus sacrificing, to a great extent, the advantage of keeping the second line out of fire, in order to secure the more important one of concentration of force. But this was only his formation for defence; for, in the same battle, his formations for attack were always in close columns.

14. Our present Infantry Tactics have adopted two new expedients to accelerate the advance of battalions, and diminish the loss to which columns of attack are liable--Division Columns and Advancing by the Flank of Subdivisions.

As Division Columns break the battalion line into several columns, each of two or three subdivisions deep, as a substitute for a single column four or five subdivisions deep, they undoubtedly diminish the loss from the enemy's artillery fire in corresponding proportion. But in compensation for this partial advantage, they have three defects:--

(1.) In moving rapidly for any distance, especially over broken or obstructed ground, both the alignment and the proper intervals between the columns will usually be lost; thus causing, in the deployment, a dangerous loss of time in re-establishing the alignment and the correct intervals.

(2.) In advancing in line of division columns, there is no means of forming square, except by passing through an intermediate formation.

(3.) The intervals between the columns are so many gaps, through which cavalry could easily penetrate, and take the columns in rear.

The line of division columns appears to have been first suggested by Marshal Marmont, who was a good artillery commander, but not necessarily, for that reason, a weighty authority on a point of Infantry Tactics.

15. The manoeuvre of Advancing by the Flank of Subdivisions is obnoxious to all the objections just pointed out in regard to Division Columns. On being threatened by cavalry, though the troops would have no intermediate formation to pass through to prepare for forming square, they would have to face into column and close to half distance, which there would often not be time to do.

In addition to this, the flank march being habitually by fours, the subdivisions would offer a tolerable mark for the enemy's artillery, and thus be exposed to a destructive enfilade.

And in forming into line, where the leading guides have not accurately preserved both their alignment and their intervals, which must be the usual case in the field, there must be more or less delay and confusion, of which a prompt and active enemy would not fail to take fatal advantage.

The mode prescribed by the Tactics (Par. 150, School of the Battalion), for executing the manoeuvre of forming line while advancing by subdivision flanks, seems also to call for remark; it being "by company (or division) into line." In other words, each individual soldier brings a shoulder forward, breaks off from his comrades, and hurries up, not on a line with them, but detached from them, and moving independently, to find his proper place. This destroys for the time being, and at a critical moment, the unity of the subdivisions, and so impairs the confidence soldiers derive from realizing that they form part of a compact mass. In thus executing this manoeuvre under fire, and near the enemy, there is danger of the men becoming confused and bewildered. For this reason, a better method of forming line would seem to be to re-form the column by a simple facing, and then to wheel into line by subdivisions.

16. The worst possible order of marching in battle, for any considerable number of men, as a battalion, for instance, is by the flank. Such a line, advancing in what is really a column of fours, would be rolled up and crushed, on the enemy's attacking its head; and would, meanwhile, be exposed to enfilade. Marching to a flank, it would be running the gauntlet of the enemy's batteries and musketry fire. In forming into line in either case, much time would be lost; as in flank marching in the field, especially when the ground is ragged or obstructed, distances cannot be preserved.

It may be here remarked, that marching to a flank in column also, whether by division, company, or platoon, is highly objectionable, as it constantly exposes the column to an enfilading fire, as well as to be suddenly charged in flank by cavalry.

III.--The Attack, how made.

1. The speed of a column of attack must never be checked for a moment, to enable it to reply to the enemy's fire. The fire of the column will be ineffective, for it will be the fire of excited men, and very limited in extent, as it can proceed from the leading division only; and the fire once begun, it will be hard to stop it. If, in order to fire, we halt the column, re-forming it under the excitement of the fire will be very difficult; and the enemy's least forward movement may then cause a rout.

At Maida, in Calabria, in 1806, the French columns attacked the English under General Stuart. When within thirty paces, the English gave them a volley. The French, stunned, as it were, began, at once, to deploy. The English fired again, and the French retreated.

At Waterloo, in the last grand attack by the French, the advance column of the Imperial Guard was decisively repulsed by the British Guards. These had been lying on the ground behind the crest of the slope until the French appeared, when they suddenly rose up and poured in a murderous volley at short range. Instead of instantly charging with the bayonet, the French hesitated, then began to deploy. The British charged at once, and drove them down the hill.

2. This dangerous halt and deployment is apt also to occur when the column finds sheltering objects by the way. Therefore, hurry by these, and hasten the step.

3. It will also tend to prevent such an untoward accident, if we furnish the columns of attack, where several are employed, with skirmishers in their intervals, as well as on their outer flanks, to draw the enemy's fire. Otherwise, the column fired into will be apt, in order to return the fire, to halt instinctively and deploy into line, which breaks up the attack.

4. From this it appears that the limited fire of a column of attack is, in fact, no defect, the highest offensive power of infantry being in the bayonet. Fire, in the attack, is generally ineffective, and sometimes injurious. It should rarely be used till the enemy has turned his back.

5. As to attacking cavalry:

Infantry may advance in line and attack cavalry safely, provided its flanks are protected. Before a long line of infantry, cavalry must retreat, or be destroyed by its fire. In the Austrian service it is said to be a received maxim, that horses will not stand before the steady approach of a mass of infantry, with bayonets at the charge, but will always retire before the infantry closes on them.

6. So, infantry in column, either closed in mass, or at half distance, may attack cavalry successfully; taking care to be ready to form square, or "column against cavalry," at the first symptom of their preparing to charge.

7. As to attacking artillery:

Before charging, the infantry sometimes first seeks the shelter of ground, using its sharpshooters to annoy it, and, if possible, to silence its fire.

Or, when circumstances are favorable, as when it can get a position near its flank, it attacks it vigorously, at once, with fire and bayonet.

But when infantry has to advance to the attack of a battery in front, it should never be in any compact formation, but always deployed as skirmishers. Otherwise, it would usually meet with a bloody repulse; especially where any considerable space of ground is to be cleared.

At the battle of Malvern Hill, the rebel General Magruder's division was sent, either in column or in line, to charge a powerful Union battery just beyond an open field a mile and three-quarters in length. The rebels rushed into the field at a full run, but encountered a murderous fire from the guns they were sent to attack, which mowed them down by hundreds. By the time they had cleared two-thirds of the ground, the carnage was so dreadful as to drive them back to the woods from which they had started. Twice more they were sent forward in the same manner, but with the same result; when the undertaking was abandoned.

8. In attacking a battery, we may often secure its capture by a volley aimed at the horses; the effect of which may prevent the enemy from carrying it off. But this should be avoided when there is a good prospect of capturing the

battery without disabling the horses; since then, if we succeed, we shall be able to immediately use the battery against the enemy ourselves.

9. In the French Revolution, the Chouans of La Vend閑 attacked the Republican batteries in several single files, of one or two hundred men each, at intervals of fifty paces. Such a formation protects the attacking columns, to a great extent, from the enemy's fire, but exposes them to destruction by a charge from the battery supports. In the absence of these, it would often be very advantageous; since, by proper drilling, these columns in one rank could be made, on arriving near the enemy, to rapidly double in two or four ranks, without halting, and then, by filing to a flank and facing, to advance by the front in a compact line.

The same formation would be useful for troops advancing to assault an intrenchment; but, as in the case of a battery, subject to the risk of being destroyed by a sudden sortie from the work.

10. Artillery is never without supports. One part of the infantry, therefore, deployed as skirmishers, should attack the guns, circling round them, and opening fire on the men and horses; while the other part attacks the support in flank. On getting sufficiently near, the assailants should try to draw the fire of the guns, and then rush on them before they have time to reload.

If a battery gets into confusion, or there is any delay in unlimbering or limbering up, then is the most favorable time to capture it by a vigorous charge with the bayonet.

IV.--Bayonet Charges.

1. When made resolutely, and without slackening the gait, bayonet charges have succeeded in nine cases out of ten.

2. The bayonet is usually more effective than grape, canister, or bullets.

At the battle of Leipsic, in 1813, Kleist's Prussian division was sent to carry the position of Probstheyda. For this purpose it was necessary to advance up a long slope, the crest of which was occupied by Drouot's artillery. The French allowed the Prussians to approach to within a short distance, and then poured into them a most destructive shower of grape, which drove them back for a moment in confusion. But they immediately rallied, and rushed desperately on again. Marshal Victor then charged them with the bayonet, and completely repulsed them.

Afterwards, having been re-enforced by Wittgenstein's Russian division, they again advanced, under a constant shower of grape from Drouot. They, nevertheless, kept advancing; and, in spite of the great loss they suffered, were about carrying the position, when the French again charged with the bayonet, forcing them down to the very foot of the declivity; where, being once more covered with grape, their repulse was complete and final.

So, at the battle of Mill Springs, in January, 1862, after the combatants had been exchanging musketry fires for several hours without any decisive result, the rebels' left was vigorously charged by the Ninth Ohio with the bayonet. This charge broke the enemy's flank. His whole line gave way in confusion, and the battle was won.

So, at Malvern Hill, in 1862, in several instances, columns of rebels whom a storm of canister and shell had failed to repulse, were driven back and routed by a dash with the bayonet, after a volley poured in at a few yards from the muzzles of the guns.

So, at the battle of Seven Pines, according to General Heintzelman's report, whenever our troops used the bayonet, their loss was comparatively light, and the enemy was driven back, suffering heavily.

3. The bayonet charge, when made from any considerable distance, should be in column; the only formation in which order can be combined with sufficient speed. But, at a short distance, a bayonet charge by a line, instantly

after firing a volley to repel an attack, will be very effective, and usually successful.

4. In ordinary cases, the charge should be prepared by first shattering the hostile masses, or, at least, wearying and demoralizing them by artillery, or by skirmishers' fire.

5. The more vigorous and resolute the charge, the greater the chance of success. The enemy never retires before a moderate advance.

6. Where the enemy is forced into a defile, a charge with the bayonet, preceded by a few rounds of grape, will complete his destruction.

7. When the enemy is behind cover, the best way to drive him from it is with the bayonet. This will cause less loss of life than to attempt to return his fire. But, in such case, the charge should be prepared, when possible, by a few shells, or rounds of canister.

8. Shots up or down a declivity usually miss. A height should, therefore, be carried with the bayonet, without firing.

The moral effect, moreover, of a steady charge of infantry up a hill, without stopping to fire, is very great; and such a charge is usually successful. Prince Czartoryski, Alexander's most experienced general at Austerlitz, admitted that he lost all confidence in the result on seeing the French infantry ascending the plateau of Pratzen, the key to the Allies' position, with a firm and decided step, without once stopping to fire.

So, at Chattanooga, in November, 1863, Thomas's troops carried the height of Missionary Ridge by a similar steady and determined ascent, in spite of the volleys of grape and canister from nearly thirty pieces of artillery, and of musketry from the rebels' rifle-pits at the summit. General Grant attributed the small number of casualties our troops sustained in the attack to the rebels' surprise at its audacity, causing "confusion and purposeless aiming of their pieces."

V.--Defence against Infantry.

1. The defence of infantry is by its fire, and therefore its proper defensive formation is in deployed lines.

2. Avoid a premature commencement of the fire. Long firing exhausts the men's energy, expends the ammunition, fouls the pieces, destroys the soldier's confidence in his weapon, and emboldens the enemy.

3. So, a fire upon an enemy while under cover, as in a wood, would be virtually thrown away. If his fire from such a position causes us any loss, he had better be shelled, or driven away by skirmishers, according to circumstances.

4. The practice of hostile regiments exchanging for a considerable time a musketry fire at a distance, is highly objectionable, as it causes a great sacrifice of life without corresponding results. Instead of standing in line for ten minutes, receiving and returning fire at a distance of three hundred yards, it would be much better to clear this space at double quick in two or three minutes, and close with the enemy; for, in returning his fire, we can do him no more harm than we receive, while nothing decisive is accomplished. The case is, of course, different where our own troops are behind cover, while the enemy's are exposed.

5. But in special cases, as where we have to cover a flank movement of our second line, or of the reserve, or to await a force coming to our support, it may be necessary to keep up an incessant fusillade, without regard to losses received.

6. Fire in action is of two kinds: the fire at will, and the fire by volleys; the former kind being the rule, the latter the exception. Although the fire at will is the one principally used, there are very strong objections to it.

(1.) The men load and fire as individuals, and generally with great rapidity, and under more or less excitement, rarely stopping to take a deliberate aim. The consequence is, that very few shots take effect, and the fire is, for the greater part, wasted, as is shown by the well-established fact that, in every engagement, for every man killed or disabled, there have been from three to ten thousand musket or rifle bullets fired.

(2.) Except on windy days, a cloud of smoke soon collects in front of a line firing at will, hiding, more or less completely, the enemy from view. The fire being then at random, it is, of course, unreliable.

(3.) The fire at will leads to a rapid and enormous consumption of ammunition. To show how serious is this objection also, it is only necessary to consider in how many instances victory has been turned into defeat by the premature exhaustion, by one or more regiments, of their ammunition.

(4.) As a necessary consequence of this rapid consumption of ammunition, the pieces soon become fouled, and thus, to a great extent, useless.

(5.) Troops under a musketry fire at will, soon become accustomed to it, and its incessant din produces on them a stunning effect, which deadens, in no small degree, their sensibility to danger.

7. On the other hand, volley firing has often been attended with decisive results, especially when it has been reserved to the proper moment, and delivered at short range. Instances of this have occurred in almost every great battle we read of in history, as also in the late War of the Rebellion. For example: at the battle of South Mountain, Doubleday's brigade was engaged with a heavy force of rebels at some thirty or forty paces in its front. Our men were behind a fence, firing at will; but their fire made little or no impression on the enemy, who attempted to charge at the least cessation of the fire. Our troops were then made to cease firing, to lie down behind the fence, and, on the enemy's approach to within fifteen paces, to spring up and pour in a

volley. This was so deadly, that the rebels fled in disorder, leaving their dead and wounded, and could not be rallied again.

At Chickamauga, in 1863, the regiments of Hazen's brigade fired only by volleys; every one of which, it is officially reported, was powerfully effective in checking the enemy's attacks.

8. Nevertheless, it has been a common military saying, and supported even by high authority, that the fire at will is the only one possible in action. This assertion implies that the rank and file are not sufficiently cool to reserve their fire, and that they must be kept constantly occupied by the excitement, noise, and smoke of their own fire, in order to make them remain steady in their ranks under that of the enemy.

As applied to raw, undisciplined, or demoralized troops, the proposition may be, to a great extent, true. But in reference to disciplined or veteran troops, whose morale has not been impaired, it will be found disproved on almost every page of military history; from which a few examples will be cited hereafter. For the present, one instance will suffice; that of Colonel Willich's regiment of Thirty-second Indiana Volunteers, at the battle of Shiloh, in April, 1862. While under fire, their commander, perceiving their own fire to have become "a little wild," caused them to cease firing, and then drilled them in the manual of arms, which they went through as if on parade; after which, they again opened on the enemy a fire, which is reported to have been "deliberate, steady, and effective."

It may be here observed that, whenever troops lose their presence of mind, there is no surer way of restoring it than by the repetition, by their officers, in their usual tone, of any words of command they have learned instinctively to obey on the drill-ground.

9. Infantry, when charged in position, should reserve its fire till it can be made with deadly effect, as at the distance of fifty paces; and the volleys should be instantly followed up by a countercharge with the bayonet on the

charging enemy. For, if our fire has staggered him, a vigorous charge will complete his repulse; and if it has not, our only chance of success is in suddenly taking the offensive ourselves.

Whilst awaiting his charge, we shall incur but little, if any, loss from the enemy's fire; for the fire of troops advancing to attack is usually of very little account.

The only disadvantage attending a volley just before we charge is, that, as the smoke veils us from the enemy's view, it will rob us, to some extent, of the moral effect of our swift advance.

But, in many cases, if the enemy see us awaiting his bayonet attack, and reserving our fire to the very last, he loses resolution, relaxes his speed, and then stops short, or retires.

At Cowpens, Colonel Howard broke and routed the British line which was advancing to attack him, by reserving his fire to within thirty yards, and then charging with the bayonet.

At the battle of Friedland, the Russian Imperial Guard charged on Dupont's division with the bayonet. The French did not wait for them to close, but rushed on with the bayonet themselves, and completely routed them.

10. A volley concentrated upon the enemy's regimental colors will usually disable the color-guard and the men near it; and, if promptly followed up by a charge, may enable us to capture the colors. This is always an important advantage; for, by the loss of its colors, a regiment is not only dispirited, but in danger of disorganization; these being its proper rallying-point.

11. When infantry is acting as a support to artillery which is attacked, it should throw out sharpshooters to reply to the enemy's skirmishers that are firing at the gunners and horses, whilst it engages the compact mass by which it is itself attacked.

If the enemy should commit the blunder of attacking the battery with his entire force, without detaching to engage the support, we should profit by it by instantly charging him in flank; but taking care not to be led away to any distance from the battery we are protecting.

12. Infantry, surrounded by the enemy, will often be able to cut its way through and escape. For this purpose, as the highest degree of concentration is required, its formation should be in close column.

VI.--Defence against Artillery.

1. The best defence of infantry against artillery is by the fire of sharpshooters deployed as skirmishers, to pick off the gunners and the horses; the main body, meanwhile, occupying the most sheltered locality it can find.

2. Where no shelter is afforded by any natural obstacles, or by irregularities of ground, it may be sometimes necessary to make the men lie down.

But this expedient should be used as rarely as possible, on account of its demoralizing tendency. Troops that have become accustomed to it cannot be expected to bravely face the enemy; and the habit is very rapidly formed. At Bull Run, in July, 1861, a whole company was seen to grovel in the dust at the mere snapping of a percussion-cap of one of their own muskets.

This demoralizing tendency does not exist, however, where troops lie down only to enable their own artillery to fire over them. This was shown at the battle of Pea Ridge, where several of our regiments lay on the ground for two hours or more, while thirty of our guns were firing over them. When, at last, this fire had silenced the enemy's guns, our infantry then rose, charged him in a compact line, and drove him from the field.

3. A line of infantry may avoid cannon-shot by advancing or retiring fifty paces. A column or a square would have to move this distance, or more,

according to its depth.

Ricochet shots may be avoided by moving fifty paces to the right or left.

This shifting of position is but a temporary expedient, it is true, for the enemy's guns will soon obtain the exact range again. But for this, several trial-shots will be requisite, thus making the enemy lose time; and, in battle, a few minutes lost or gained have often decided between victory and defeat.

4. When the enemy opens an artillery fire on a square, preparatory to a cavalry charge, his fire must cease when his cavalry approaches the square; say, on its arriving within one hundred and fifty yards. To avoid the artillery fire, the square may safely remain lying down until the hostile cavalry has reached this point. For, as they will require about half a minute to clear the intervening ground, the square will still have time enough left to rise, align its ranks, and deliver a volley before the cavalry reaches it.

VII.--Defence against Cavalry.

1. The discipline of infantry is never put to a severer test than when it is required to resist a charge of cavalry, properly made. The moral effect of a charge of a body of horse at full speed, on the troops waiting to receive it, is like that caused by the swift approach of a locomotive under full steam, seeming quite as irresistible. It would be so in reality, but for the counter effect produced both on the horses and their riders by the sight of the infantry standing firm and reserving its fire. I have been told by an old cuirassier officer, who served through the campaigns of Napoleon with distinguished bravery, that there was no operation that his regiment so much dreaded as a charge upon well-disciplined infantry.

2. This counter moral effect on the charging cavalry is the greater, the longer the infantry reserve their fire; since, the less the distance at which it is delivered, the more fatal will be its effects. A volley at long range is not destructive enough to check the cavalry's advance; while this effect has often

been produced by the infantry merely withholding its fire till the cavalry has approached very near; and a volley delivered at the very last moment has, in by far the greater number of instances, effectually repulsed the charge.

Infantry should, therefore, let cavalry approach to within forty paces, or nearer still, and then give them a general volley.

At the battle of Neerwinden, in 1793, the Austrian cavalry was repulsed by the French infantry under Dumouriez, by a volley poured in at the very muzzles of the pieces.

At Austerlitz, a Russian cavalry charge on French infantry in line was repulsed by a volley delivered so near, that it stretched four hundred troopers on the ground. The rest dispersed in disorder to the right and left.

3. The armor of cuirassiers is bullet-proof. To repel a charge of these troops, therefore, it will be necessary to aim at the horses. Their armor is so heavy, that the mere fall of the riders on the ground is usually sufficient to disable them, as was the case with the French cuirassiers at Waterloo.

4. Infantry in line, in two ranks even, may withstand cavalry, if in compact order, and attacked in front. But the slightest cavalry charge on the flank of a line will rout it.

At Quatre Bras, a French infantry line, advancing, repulsed a charge of the Brunswicker Lancers under the Duke of Brunswick, by receiving it in steadiness and good order, and then pouring in a destructive fire.

But, in the same battle, the Sixty-ninth British Regiment was instantly rolled up and destroyed by a charge of French cuirassiers on its flank.

5. Where infantry is well disciplined, and its commander is cool and prompt, it may sometimes avoid the effect of a cavalry charge by other means than its fire, or formation in square. At Talavera, a French infantry division, drawn up

in close column, seeing an English cavalry regiment charging down upon them, avoided the shock by simply stepping aside, thus allowing the cavalry to pass by them. A portion of the charging troops wheeled round to follow them; but, by the cross-fire of another division, and the charge of other cavalry, which fell upon it in its confusion, it was completely annihilated.

6. A line of infantry charged by cavalry in flank, and so suddenly as to allow no time to form square, could hardly escape destruction. It would seem that the best course to be adopted in such a case would be to open the ranks by a rapid and simultaneous movement of both of them, thus compelling the charging cavalry to ride between them. If the front rank should then face about, this would bring the cavalry between two fires, which might be poured in with most destructive effect.

But where the cavalry charges with a very wide front, or in line, this manoeuvre might be difficult, or impossible.

7. Whenever an infantry line is charged by cavalry in front, and it is doubtful whether it will stand the shock, the wisest course would seem to be to make the men lie down, and let the charging cavalry leap over them. This the horses will instinctively do, with but little risk of injury to the men, provided they lie in a position parallel to the line of battle, thus presenting the least possible depth. It is said that the British infantry has sometimes done this, and risen up again immediately after the cavalry had passed. The cavalry could thus be promptly taken in rear.

8. In retreating, when threatened by cavalry, if there be a long plain in our rear, we must retire slowly. But if cover, or ground unfavorable to cavalry, be near, we must reach it as soon as possible.

VIII.--Squares.

1. In 1813, France was nearly exhausted of soldiers, so that Napoleon, on hastily preparing for his campaign of that year, was obliged to incorporate

into his army a large number of raw conscripts, who had scarcely begun their elementary drill. On the route to their respective points of concentration, he accordingly ordered his columns to halt each day, to practise the three movements which he considered to be the most important for infantry to be familiar with. These were, forming battalions in square, deploying in line, and re-forming in column of attack.

2. In the Austrian service, squares formed by a column in mass are considered preferable to hollow ones, on the supposition that though horses will recoil from a dense mass, they may be easily brought to break through a shallow formation, over which they can see the open ground. But this theory seems to be refuted by numerous facts. A large proportion of the formations that have successfully repulsed cavalry, since the beginning of this century, have been hollow squares.

3. The rule laid down in the Tactics (Par. 143, Skirmishers), directing the skirmishers, in rallying on the square, to "come to a ready without command, and fire upon the enemy; which will also be done by the reserve, as soon as it is unmasked by the skirmishers," is an unsound one, for a compliance with it would be dangerous. A square cannot expect to repulse cavalry by an irregular fire at will, but only by well-directed volleys. If cavalry charge a square firing irregularly, it will probably rout it. On the other hand, if a square wait coolly till the cavalry is at twenty paces, its volley will be murderous. At Waterloo, the Allied squares that reserved their fire till the French cavalry had arrived at from twenty to forty paces, invariably repulsed it. At that battle, Ney led eleven cavalry charges against the British squares, every one of which failed.

At the opening of the campaign of 1813, Napoleon had, comparatively, but a handful of cavalry; so few, that they had to keep close to their infantry for protection. In crossing the plains of Lutzen, a large and splendid cavalry force of the Allies, supported by infantry and by horse-artillery, made an attack on Ney's corps, which consisted chiefly of young and raw recruits, who saw an enemy for the first time. The situation was extremely dangerous, and Ney and his principal generals threw themselves into the squares to encourage them. By volleys delivered at a signal, the enemy's charges were all repulsed, and the conscripts acquired great confidence from the ease with which this was done. Ney then broke up his squares, and, pursuing the enemy in columns, completed their repulse.

At Auerstadt, in 1806, Davoust's French squares had to sustain a long succession of charges from ten thousand Prussian horse. By reserving their fire, each time, to within thirty or forty paces, its effect was so deadly, that a rampart of dead and disabled men and horses was soon formed around the squares, and the charges were all repulsed.

So, at Jena, on the same day, Ney, posted in a square, allowed the Prussian cuirassiers to charge up to within fifteen or twenty paces, when the front attacked, at his word of command, poured in a fire which completely repulsed the charge, strewing the whole ground with dead and wounded. The Prussian cavalry, in that battle, are said to have been "terrified at the sight of a motionless infantry reserving its fire."

Again, at Mount Tabor, in 1798, General Kleber, marching with an infantry division of only three thousand men, over an immense sandy plain, was attacked by twelve thousand Turkish horse. The French squares resisted their successive charges for six hours, by means of volleys reserved till the enemy were at the very muzzles of their guns; which soon built up a rampart around them of men and horses. Bonaparte then arrived with another division. Dividing it into two squares, he rapidly advanced them in such a manner as to enclose the Turks in a kind of triangle; when, by a sudden fire upon them from three points at once, he drove them upon each other in confusion, making them flee in every direction.

It may be observed, that advancing or manoeuvring in squares is practicable only on open and level plains, like the sandy deserts of Egypt and Syria.

4. The best reliance of an infantry square being, therefore, on its fire by

volleys, the men should be instructed to come to a charge bayonet, instead of a "ready," immediately on forming square. From this latter position, there would be much greater danger of the volley being prematurely delivered. The fire of a single excited man will usually be followed by a general discharge.

5. It may be often advisable that the volley should be delivered by both ranks at once, and not by a single one. Par. 1191, School of the Battalion, directing that "a battalion, in square, will never use any other than the fire by file, or by rank," should therefore be amended.

6. Moreover, in view of what has been said as to volley-firing, and of the examples that have been cited in confirmation, there is reason to doubt the wisdom of the direction contained in Par. 67, School of the Company: "The fire by file being that which is most frequently used against an enemy, it is highly important that it be rendered perfectly familiar to the troops. The instructor will, therefore, give it almost exclusive preference."

The fire by file, after its commencement, becomes a mere individual fire at will. Independently of the general ineffectiveness of this kind of fire, one would have supposed that the instructor's attention should be rather directed to accustoming the men to the more difficult reserved fire by volleys, instead of practising them almost exclusively in a fire which, once learned, they will use instinctively, and without any practice at all.

7. Infantry breech-loading weapons would be very useful to troops in square, when charged by cavalry; since, being rapidly reloaded, they would enable the square to repulse, with a volley, each subdivision successively, where the charging column is formed at the usual distances. But it is doubtful whether, on the whole, these weapons are preferable to muzzle-loaders. Certain it is, that they exhaust the ammunition much more rapidly, and so cause a suspension of fire, and a withdrawal from the line of battle, till a new supply can arrive. And, to obtain this new supply, a long time is generally required; infantry ammunition being usually carried in the second, or more distant ammunition train, instead of the first, or nearest one, as it ought to be.

8. Although a reserved fire is much the most reliable in repulsing cavalry, the men may sometimes be ordered to commence the fire at a considerable distance. In such case, they should be instructed to aim at the horses, instead of their riders, as affording a better mark.

9. European cavalry is often practised, on arriving within four hundred yards, or effective grape-shot distance, of an infantry square, to halt, and then open at the centre, unmasking a battery of horse-artillery, which plays for a certain time on the square, when the cavalry closes again, and charges.

A square, however, attacked in this manner, is not in so much danger of being broken as might be imagined. The enemy's guns, after being unmasked, would usually require several trial rounds to get the exact range; and our sharpshooters, who could safely be thrown forward one hundred yards, with the new rifled arms, ought, in the mean time, to inflict such loss on the cavalry, as well as on the battery, as to cause it either to retire, or to charge feebly, and, therefore, ineffectively. At the very worst, the square would have ample time to re-form its ranks, and deliver a deadly volley before the cavalry could reach it, as it also would if this operation were attempted much nearer, say at two hundred yards. In this last case, a few volleys from the square itself, with the new arms, would probably be destructive enough to prevent the charge altogether.

10. It is hardly necessary to observe, that troops formed in square, when charged by cavalry, can secure their safety only by standing firm. A single opening will suffice to let in the enemy, who will then easily ride over the square, and cut it in pieces. Whereas, if the square remain unbroken, cavalry can inflict upon it no loss, or but a trifling one.

11. In repulsing a cavalry charge, coolness and presence of mind will sometimes enable troops to accomplish extraordinary results.

At Quatre Bras, the square of the Forty-second Highlanders was not completed, the companies still running in to form the rear face, when the enemy's leading troop entered. But the square, nevertheless, finished its formation; and the French cavalry, caught, as it were, in a net, was soon destroyed by the concentrated fire of all the fronts, which had faced inward.

In the same battle, the Forty-fourth British Regiment, standing in line in two ranks, was suddenly charged in rear by the French Lancers, who had dashed round one of their flanks for that purpose. The rear rank suddenly faced about, and, at a very short distance, poured in a deadly fire, which put them into confusion. On their way back to re-form, the front rank, in its turn, gave them a volley, which destroyed great numbers of them, and completed their rout.

12. Even when a square has been actually broken, it is not necessarily lost. If the troops are brave and well disciplined, it may sometimes be rallied again, re-formed, and made to repulse the attacking cavalry, as was the case with some of the Allied squares at Waterloo.

So, at the battle of Pultusk, in 1806, a French battalion that had been broken and overthrown by Russian cavalry, immediately rallied, fell on the troopers floundering in the mud, and dispatched them.

So, at the battle of Krasnoe, in 1812, a large Russian square was retreating before the French cavalry masses. Occasionally, in order to pass a narrow defile, it was obliged, temporarily, to break the square. At these times the French made furious charges, penetrated into the column, and captured men and guns. But as soon as the defile was passed, the Russians instantly re-formed the square, and continued their retreat. They finally succeeded in reaching Korytnia, after killing and wounding some four hundred or five hundred of the French; though with the loss of eight guns, one thousand prisoners, and seven hundred or eight hundred hors de combat, out of five thousand or six thousand men.

IX.--Skirmishers.

We shall consider--

First, THEIR USE.

Secondly, HOW THEY ARE POSTED.

Thirdly, HOW THEY ARE HANDLED.

Fourthly, RULES FOR INDIVIDUAL SKIRMISHERS.

A. THEIR USE.

1. In approaching the enemy through a wooded or broken country, skirmishers thrown out in advance, and on the flanks of the leading column, are absolutely indispensable, in order to reconnoitre the ground, and prevent a surprise.

2. Skirmishers protect the main body, or any particular portion of it, from attack while manoeuvring.

A regiment, or a brigade, in covered ground, whether the enemy be visible or not, should never change its position in battle, or manoeuvre, without the protection of a skirmishing line.

3. They furnish a screen, behind which the main body may hide its movements, and be enabled to attack at an unexpected point.

4. Where a ravine, a wood, or other similar obstacle causes a break in our line of battle, by occupying it with skirmishers we guard it against penetration by the enemy, and connect the separated corps with each other.

5. Skirmishers may be used to alarm the enemy at a point where he expected no attack, and thus create a diversion.

6. By their attack at various points, they serve to unmask the enemy's position.

7. They may be employed to open the way for a charge with the bayonet.

At the battle of Stone River, the rebels, on one occasion, advanced in line, with a double column in rear of each wing, preceded by a double line of skirmishers, who reserved their fire till close to our line, when they halted, poured in a murderous fire, and fell back on their main body, which then rushed forward. Both our first and second lines, staggered by this sudden and destructive fire, were swept from the ground.

8. Skirmishers have been sometimes thrown forward to test the spirit and disposition of the enemy.

At Biberach, in 1800, the French general St. Cyr, after having carried the place, and driven the Austrians through the defile in rear of it back upon their main body, posted on the heights of Wittenburg, sent forward a strong line of skirmishers to open fire on them, with the view of ascertaining their temper and disposition after their vanguard had been defeated and driven in. This drew forth a general and continued discharge, like that which demoralized troops are apt to indulge in to keep up their spirits by their own noise. Seeing this, St. Cyr instantly prepared to charge, although he had with him but twenty thousand men, and the Austrians numbered sixty thousand, and were in a strong position. The result justified his decision; for, on the near approach of the French, the Austrians fired a volley or two and then retreated in confusion.

9. Skirmishers should accompany columns of attack; for--

(1.) They increase the confidence of the troops they accompany. Placed between the columns, they advance boldly because the columns advance, and the columns advance boldly because the skirmishers do.

(2.) Preceding the columns, by driving back the enemy's skirmishers, and diverting his fire to themselves, they keep the attacking columns as free from loss as possible till the shock.

They, moreover, serve to annoy the troops we are about to attack, by the incessant sharp buzzing of their deadly bullets among them, like so many bees, killing some and disabling others; and this, sometimes, to such a degree as to demoralize them.

It is said that, at Waterloo, the swarms of skirmishers that covered the French attacking columns so galled and excited the stationary columns and squares of some of the Allies, as to nearly drive them from the field.

(3.) On the flanks of a column, they cover them from attack.

(4.) They draw the enemy's fire prematurely, and thus render it comparatively ineffective.

(5.) They prevent the columns from halting to deploy and fire.

(6.) They may sometimes conceal the direction of the march of the attacking column, and even seize the guns that have been playing on it.

10. In defence, if they can encircle the enemy's advancing column, they may destroy it by their concentric fire.

11. In a retreat, skirmishers cover the rear, so long as the enemy attacks without cavalry.

12. The NEW RIFLED ARMS have obviously much increased the effectiveness of skirmishers.

B. HOW POSTED.

1. They should be always near enough to the main body to be supported by it, if hard pressed, and also to enable the main body to profit at once of any advantage that may have been gained by them.

2. They should cover the main body, both in front and in flank, except where the ground may render this impracticable or unnecessary; and, in defensive positions, they should occupy every point from which the enemy's skirmishers might annoy us.

3. In a defensive combat, they should be so posted as to take the enemy's attack in flank:

(1.) Because their fire will be thus the more destructive; and--

(2.) They will not be exposed to be driven back by the enemy's fire, or by his advance.

4. If thrown into an enclosure, they must have an easy exit. Skirmishers feeling themselves in danger of being cut off, will lose somewhat of that coolness which is so essential to their efficiency.

5. They should not be kept stationary behind a straight line, as a wall, a fence, or a hedge; for this would expose them to enfilade.

6. Skirmishers are only auxiliary to the main force, and are not capable, by themselves, of effecting any decisive result. Therefore, in order not to exhaust the men, heavy skirmishing lines should not be used, except to lead a decided advance, or to repel one.

7. The principle is, to post skirmishers so as to give them the maximum of shelter, whilst inflicting the maximum of loss on the enemy. This applies to the placing of the whole line, and to the separate groups. The way skirmishers produce their effect is by sharpshooting, which requires calmness;

and the more completely sheltered they are, the calmer they will be, and the more deadly will be their aim.

C. HOW HANDLED.

1. Deploy them before coming within range of musketry; for infantry in compact order is a good target for the enemy.

2. They should be kept well in hand; especially at the moment of success, when they are in danger of rushing headlong to destruction.

3. Coming upon the enemy's main body, they should occupy him in front and flank till our own main body gets up.

4. Except in urgent cases, never deploy a line of skirmishers on a run; for this makes them lose breath and calmness, and, with their calmness, their accuracy of aim.

So, after deployment, avoid all rapid and violent movements.

5. Skirmishers become exhausted after long firing. The longer they continue out, the worse they shoot. Therefore, relieve them often.

6. Skirmishers should be accustomed to lie down at a given signal; as it is sometimes very important that both our artillery and infantry should be able to fire over them.

7. In retreat, skirmishers occupy every favorable point for holding the enemy in check.

D. RULES FOR INDIVIDUAL SKIRMISHERS.

1. In advancing, in retreat, or at a halt, use every cover that presents itself.

2. Preserve the alignment and the intervals, so far as possible. On open ground, this may be done perfectly. In woods, skirmishers should never, for a moment, lose sight of each other.

3. The security of the flanks should be looked out for by the men near them.

4. Run over exposed ground as quickly as possible.

5. Approach the crest of a hill with great caution.

6. If threatened by artillery alone, advance and kill off the men and horses before they get into battery. When the pieces have got into battery, lie down, if on exposed ground, till they limber up again, and then recommence the fire.

7. A skirmisher, with the new rifled arms, ought, at five hundred yards, to be more than a match for a gun; for, in men and horses, he has a much larger target than the gun has in him.

Again, with the new rifle shells, he may be able to blow up a caisson.

8. Neither should a skirmisher have much to fear from a single horseman. With his bayonet fixed, he would usually be able to defend himself successfully against the trooper, whose sabre is the shorter weapon of the two; more especially, if he will take care to keep on the trooper's left, which is his exposed side.

9. Never lose your calmness. Your power consists, not in rapid firing, but in the accuracy of your aim. Avoid all hurried and violent movements; and never raise your gun till sure of a shot.

10. The aim, according to the Tactics, is made by bringing the gun down, instead of raising it up. However little the soldier may be excited, he will be apt to pull the trigger more or less too soon; that is, while the muzzle is yet too elevated. This is the reason why infantry missiles usually fly too high. The

difficulty would not be obviated by causing the aim to be made by raising the piece; for then the same disturbing cause already mentioned, the soldier's excitement, would make the shots fly as much too low, as they now fly too high.

Rapid firing is another cause of this incompleteness of aim. Infantry firing is already too rapid to be effective; so that what is claimed for the new breech-loading weapons as an advantage, that they increase the rapidity of fire, furnishes, on the contrary, a strong objection to them. The effectiveness of the fire of a sharp-shooter, especially, will be usually in inverse, instead of direct proportion to the number of shots he delivers in a given time.

In view of this, and of the tendency to pull the trigger before the muzzle is sufficiently depressed, it has become an established maxim, to

"Aim low, Fire slow"

TACTICAL USE OF ARTILLERY.

The subject will be treated under the following heads:--

I.--HOW POSTED WITH RESPECT TO THE GROUND. II.--HOW POSTED WITH RESPECT TO OUR OWN TROOPS. III.--HOW POSTED WITH RESPECT TO THE ENEMY. IV.--POSTING OF BATTERIES AND PIECES AS BETWEEN THEMSELVES. V.--HOW USED. VI.--ITS FIRE. VII.--ITS SUPPORTS.

I.--How posted with respect to the Ground.

1. Artillery has a much longer range than musketry. In order to avail ourselves of this advantage, we must so post it as to overlook all the ground to which its utmost range extends. It therefore requires an elevated position.

2. It has been considered an additional advantage of a commanding position for artillery, that it enables our guns to cover our infantry, attacking or

attacked, by firing over their heads.

This was done by the French at Waterloo, apparently with great effect. But the advantage is a doubtful one; for firing over our own troops, especially with cast-shot or shell, is very dangerous to them, and is apt to intimidate them. It moreover furnishes to the enemy a double target. The shot which miss our troops will be apt to fall among the guns behind them; and some of those which do not reach the guns, will probably take effect among the troops in front of them.

3. But very high points are unfavorable positions for batteries. Batteries so placed would not command the ground immediately below them; as guns cannot be depressed to fire below a certain angle without soon destroying their carriages. And this would facilitate their capture; for, once arrived on the ground near them, the assailants could not be injured by their fire. It has been estimated that the slope in front of a battery should not exceed one perpendicular to fifteen base.

4. When guns have to be used as a support to other parts of the line, which is often the case, their capture might lead to serious consequences. They should therefore have the ground clear of all obstacles which may mask their fire, not only in front, but to their right and left.

5. Although the most favorable position for guns is an eminence sloping gradually towards the enemy, an open and level plain is by no means an unfavorable one; for, on such ground, the enemy will be visible at a great distance, and our shot may act by ricochet, which causes more destruction than ordinary point-blank firing.

For ricochet, firm and even ground is requisite; on soft or rough ground it is not attainable.

6. In enfilading the enemy's position, or in raking his advancing columns from head to rear, a grazing fire is the most destructive that can be used. This

consists of a long succession of ricochets at low heights. Where the ground is level and firm, we can obtain this fire at a short distance from the enemy; as, on such ground, ricochet shots do not rise much. But where the ground is uneven, to obtain such a fire, a more distant position will be requisite.

7. Muddy ground is unfavorable for artillery. Over such ground, its carriages move slowly, and its fire is less effective. Balls cannot ricochet; and shells often sink into the mud, and thus are either extinguished or explode with but little effect.

Napoleon depended so much on his artillery at Waterloo that, although every moment was precious, he delayed commencing the battle till his chief of artillery had reported the ground, which had been covered by a soaking rain, to be sufficiently dry for the movements and effectiveness of that arm. The three hours' delay thus caused, would have sufficed him to crush Wellington's army before the arrival of the Prussians.

8. Stony ground is a bad location for a battery; for the enemy's shot will scatter the stones around it with more or less fatal effect.

9. Rough or uneven ground immediately in front of a battery is not objectionable, as it will stop the enemy's shot.

10. A battery, when it is possible to avoid it, should not be posted within musket range of woods, bushes, ravines, hedges, ditches, or other cover from which the enemy's sharpshooters might kill off the gunners, or, by a sudden dash, capture the guns.

11. To prevent the enemy from approaching a battery under cover, it should be so placed as to be able to sweep all villages, hollows, and woods, in front and in flank.

12. In taking up a position, a battery should avail itself of all inequalities of the ground, for the shelter of its pieces and gunners, or of its limbers and

caissons, at least.

For the same purpose, a battery posted on an eminence should have its pieces some ten paces behind its crest.

13. Where the ground affords no shelter, and where the position of the guns is not likely to be changed, it may be worth while to cover them by breastwork, some three feet, or more, high.

II.--How posted with respect to our own Troops.

1. In order to be ready to support the flanks of our attacking columns, and to aid in the defence in every part of the field, batteries should be placed at several different points in the line of battle.

2. In a defensive battle especially, as it is uncertain on what point the enemy will mass his principal attack, the artillery should usually be distributed through the whole line.

3. A line of battle has been compared to the front of a fortification, of which the infantry is the curtain, and the artillery batteries the bastions.

4. The lighter guns should be placed on the salient points of our line, from which they can be more easily withdrawn; the heavier guns, constituting the stationary batteries, on the more retired points.

5. Pieces should not be placed in prolongation with troops; for this would be giving the enemy a double mark. Artillery posted in front of other troops will draw a fire on them. When a battery must be placed in front of the line, let the infantry in rear of it clear the ground by ploying into double columns.

6. Never place artillery so as to impede the movements of the other two arms. A battery posted in front of the centre would often hamper the movements of the infantry; besides being peculiarly exposed to a converging

fire from the enemy's batteries.

7. The safest position for a battery is on that wing which is most secure from a flank attack.

But guns should re-enforce the weaker points, thus making the enemy attack the strongest ones.

Therefore, where a wing is weak, place the largest number of guns there, to support it. If we have one wing entirely uncovered, of four batteries, for instance, we should give three to the uncovered wing.

8. Of the heavy batteries, one, at least, should be placed in the first line, so that we may be able to open an effective fire on the enemy at the earliest possible moment.

9. The prompt use, at the proper moment, of the reserve, may decide the battle. The movements of heavy artillery, therefore, are too slow for the reserve, which should have most of the light pieces. Horse artillery is especially suitable for it.

10. Guns near an infantry square should be posted at its angles. If the square is charged by cavalry, the gunners run into the square, after filling their ammunition pouches, which they take in with them, as well as their sponges and other equipments. The limbers and caissons are sent to the rear; or, if there is no time to do this, they may be brought into the square. If this is impossible, they may be formed into a barricade.

At Waterloo, on the French cavalry's retiring from their charges on the enemy's squares, the British gunners rushed out from the squares in which they had taken refuge, and plied their guns on the retiring squadrons.

III.--How posted with respect to the Enemy.

1. If the enemy's batteries are concentrated in one position, by placing our own batteries properly we may obtain a powerful cross-fire on them.

2. It is always advantageous to so dispose our batteries as to take those of the enemy in enfilade.

At the battle of Murfreesboro', in December, 1862, a rebel battery, being taken in enfilade by one of our own, was silenced in about five minutes.

3. So, also, if we can obtain an oblique or enfilading fire on his troops, it will be very destructive. A flanking battery, raking the enemy's position, is often enough, of itself, to decide a battle.

Thus, the battle of Chippewa was finally decided by our getting a gun or two in a flanking position, enabling us to enfilade the British line.

So, at the battle of Shiloh, the rebels' triumphant advance on the evening of the first day was effectually checked by the fire of our gun-boats Tyler and Lexington, which had taken an enfilading position opposite their right flank.

4. For this reason, we must never post one of our own batteries so that the enemy's guns will take it obliquely, or in flank; unless, indeed, by doing so, we may probably obtain some important and decisive effect before it can be destroyed, or made unserviceable.

5. Batteries should be so placed as to command the whole ground in our front, even almost up to our bayonets, and so as to be able to direct their fire towards every point; at all events, so that a fire can be kept up on the enemy till he is within short musket-range.

It is manifest that the best position for a battery, to enable it to effectually cover the entire ground in our front, would not be in our line of battle, but in advance of one of its flanks, from which it could take the enemy's troops advancing over it, in enfilade.

6. Artillery fire from an unexpected quarter always has a powerful moral effect. Two guns, even, hoisted up to a place where the enemy does not dream of there being any, may have a decisive effect.

IV.--Posting of Batteries and of Pieces as between themselves.

1. The best mode of posting batteries is in the form of a crescent, its horns pointing towards the enemy, or forming the sides of a re-entering angle; for this gives a convergent fire to the enemy's divergent one.

Its inconvenience is, the exposure of its flanks to attack, or to enfilade. Therefore, when such a position is adopted, its flanks must be protected by natural obstacles or by artificial defences.

2. Batteries, or parts of batteries, should be at supporting distances from each other; that is, not over six hundred yards apart, so as to effectually cover the whole ground between them, in case of need, with grape and canister. When rifled guns are used, this distance may be increased.

3. A long line of guns in our line of battle is objectionable; for, if it should become necessary to withdraw them, they would leave a dangerous interval.

4. It is dangerous to collect a great many pieces in one battery, especially in the beginning of an action, when the enemy is fresh, for it strongly tempts him to capture it. When used, such a battery should have powerful supports to protect it, or should be sheltered by a village, a defile, or other cover, occupied beforehand.

5. Although, to be used offensively, guns should be in strong masses, in order to strike a decisive blow on some single point; this is by no means the case when used defensively; for,

(1.) It is only when guns are more or less scattered over different parts of the field, that they can be made to give a cross-fire on the enemy's advancing columns, or on any part of his line.

(2.) If the position where they are massed does not happen to be attacked, they become useless, while stripping the rest of the line.

(3.) If they are captured, all the artillery is lost at once, as happened to the Austrians at the battle of Leuthen, causing their defeat.

6. A certain number of pieces of horse-artillery must always be kept in reserve, so that, if an artillery fire at any point should be suddenly wanted, it may be furnished with the least possible delay.

7. Guns of various calibres should never be in the same battery, to prevent confusion as to the respective ranges, and in the supply of their ammunition.

8. An independent section or battery should never consist of howitzers alone, for the proper fire of these pieces is too slow to be effective in repulsing an attack on them.

9. There should always be wide intervals between the pieces; otherwise the battery would offer too good a mark to the enemy.

V.--How used.

First, GENERALLY. Secondly, IN OFFENSIVE COMBAT. Thirdly, IN DEFENSIVE COMBAT. Fourthly, AGAINST INFANTRY. Fifthly, AGAINST CAVALRY. Lastly, AGAINST ARTILLERY.

A. GENERALLY.

1. So far as is possible, guns should be kept hidden from the enemy till the moment of opening on him. They may be masked by the ground, or other cover, natural or artificial, or by troops placed in front of them. The surprise

will add much to their effect. Moreover, concealed, they will be less exposed to be taken. Nothing discourages troops more than the loss of a battery at the beginning of an action.

2. A desultory and indiscriminate artillery fire will accomplish nothing. To effect any thing important, it must be concentrated on some object; and the fire must be persevered in till the desired effect has been produced.

3. It is a general principle that artillery should not reply to the enemy's batteries, unless compelled to by their effect on our own troops. To obtain the most decisive effects from artillery fire, it should be directed on the enemy's troops, instead of his guns.

4. If it should become advisable to silence one of his batteries, it will be done more promptly and effectually by the employment, for this purpose, of two of our own batteries, than of a single one.

5. There is usually great advantage in keeping our batteries constantly shifting their position; for then--

(1.) They have the effect of a surprise, by opening on the enemy at some unexpected point.

(2.) They make the enemy believe our guns to be more numerous than they really are.

(3.) They are in less danger of being captured.

But these changes of position are attended with this inconvenience, that they expose the horses to be taken in flank by the enemy's batteries and sharpshooters.

6. The movements of a battery in the field should be as rapid as possible; for, while moving, it is helpless and exposed.

Moreover, celerity of movement and accuracy of fire will often more than compensate for inferiority in the number of guns; as was the case at the battle of Palo Alto, in the Mexican War, where the enemy's guns outnumbered ours two to one.

B. IN OFFENSIVE COMBAT.

1. When used to prepare for an attack of infantry or cavalry, artillery concentrates as much fire as possible on the point where the attack is to be made, in order to overcome the resistance there, and thus make success easy.

2. When there are several points on which our fire should be directed, we must not batter them all at once, but concentrate our whole fire on them in succession.

3. In attack, artillery should not be split up among different brigades or divisions; else no decisive result can be expected from it. Whole batteries, used together, will have a more telling effect than if scattered over the field in separate sections.

In no case should less than two pieces be used together; for, while one piece is being loaded, the piece and its gunners need the protection of another one ready to be discharged.

4. Pieces in support of an infantry column of attack should never be in its rear, but on its flanks, near its head, in which position it will best encourage the infantry. But if a battery have already a position from which it can afford to the attack effective assistance, it should remain in it; sending a few pieces to accompany the infantry, which always greatly values artillery support.

5. Powerful effects may be produced by the sudden assemblage of a great number of guns on some particular point. This was a favorite manoeuvre of Napoleon; who, by his rapid concentration of immense batteries of light

artillery on the important point, usually obtained the most decisive results. At Wagram, for instance, when Macdonald's column was ready to make its great charge on the Austrian centre, Napoleon suddenly massed one hundred guns in front of his own centre, and made it advance in double column at a trot, then deploy into line on the leading section, and concentrate its fire on the villages forming the keys to the enemy's position, in front of his right and left wings respectively; each battery opening its fire on arriving at half-range distance. The effect was overwhelming.

6. The nearer artillery delivers its fire, the more powerful, of course, are its effects. Horse artillery, in sufficient strength, attacking the enemy at short grape-shot distance, say within three hundred or four hundred yards, may lose half its pieces, but with the other half it will probably decide the battle at that point.

At Palo Alto, Duncan's rapid closing with his guns to less than half range, drove back the Mexican right wing, which could not stand the destructive fire.

7. Horse artillery does not usually attempt to follow up cavalry in its attack; but takes a position to cover its retreat, if repulsed, or to push forward in support, in case of success.

8. When cavalry has to debouch from a defile, horse artillery may render it most effectual assistance, by taking a position that will enable the cavalry to form without fear of being charged and destroyed while forming.

C. IN DEFENSIVE COMBAT.

1. Artillery should always reserve its fire till the enemy's real attack.

2. It should play on that portion of the hostile force that threatens us most.

3. It should wait till the enemy has come within destructive distance, and then open on his columns with a concentrated fire.

4. It should protect our troops while manoeuvring, and accompany them in retreat.

5. We must subdivide our batteries whenever we wish to obtain cross-fires on the head of an advancing column, or on the ground in front of a weak part of our line. By so doing, we compel the enemy to divide his own artillery in order to reply to our fire.

6. A sudden concentration of a great number of guns at some particular point may be used with the same decisive effect in a defensive, as in an offensive battle; though in this case, artillery plays, for the time being, a part strictly offensive.

At the battle of Friedland, where the French were attacked by the Russians in overwhelming numbers, Ney's corps was driven back by a terrific concentrated fire, in front and in flank, from the Russian batteries on the opposite side of the river; its own artillery being too feeble to stand before them. Seeing this, Napoleon instantly ordered all the guns of the different divisions of the corps next to Ney's, on the left, to be united and thrown in one mass in front of Ney's corps. Taking post at some hundred paces in front, these batteries, by their powerful fire, soon silenced the Russian batteries; then advancing on the Russian troops that had crossed the river to within grape-shot range, they made frightful havoc in their deep masses. The French infantry, profiting by this, rushed forward and captured the village of Friedland, driving the enemy in their front over the bridges, which they then burned. This was decisive of the battle; for the whole Russian army was then driven into the river.

So, at the battle of Kunersdorff, in 1759, after Frederick's left and centre had driven the Russians, and captured seventy guns and many prisoners, Soltikoff promptly massed the whole artillery of his right wing at a single point behind a ravine, which, by its concentrated fire, swept away the flower of the Prussian army in their efforts to force its passage; and Frederick was badly defeated.

7. When compelled to retreat, guns should retire successively, in echelons of batteries, half-batteries, or sections, in order that the fire of one may cover the limbering up and retreat of another.

Besides the mutual support thereby afforded, these successive face-abouts of artillery have a powerful moral effect on the pursuing enemy, already more or less disorganized by success.

It would be well for some pieces in each echelon to be kept loaded with canister, so as to drive the enemy back if he should press very close.

8. The protection afforded by artillery in retreat is very powerful, as it keeps the enemy constantly at a distance.

A fine example of the use of it for this purpose occurred just before the battle of Pea Ridge, in March, 1862. A rear-guard of six hundred men, under General Sigel, was retreating before a force of four regiments of infantry and cavalry, that followed and attacked it on every side. Sigel disposed his guns in echelons, the one of which nearest the enemy played on his attacking squadrons with grape and shell, which suddenly checked them. Instantly profiting by their temporary hesitation, the echelon limbered up and galloped away to another position, while the next echelon, again checking the enemy by its fire, followed its example. By this means, Sigel, cutting his way through a vastly superior force, succeeded, after a retreat of ten miles, in rejoining the main body with but trifling loss.

9. If driven to the last extremity, the gunners should try, at least, to save the horses, and to blow up the caissons they have to abandon.

At the battle of La Rothie, in 1814, where Napoleon, with thirty-two thousand men, was overwhelmed by one hundred thousand Allies, and was obliged to leave fifty guns on the field, he succeeded in bringing off all his gunners and horses.

D. AGAINST INFANTRY.

1. It is an important rule that artillery should play on the enemy's troops, without attending to his batteries, except in urgent cases.

2. Against a deployed line, whether marching by the front or by a flank, case-shot, that is, grape, canister, and spherical case (sometimes called shrapnel), are most suitable; as these all scatter, right and left, to a considerable distance.

The best effect of canister is within two hundred yards. Beyond three hundred and fifty yards, it should not be used.

The best effect of grape is within four hundred yards. Over six hundred yards, it is not very effective.

Spherical case is effective at much greater distances, its range being nearly equal to that of solid shot.

Against a line of infantry at a greater distance than six hundred yards, spherical case should be used, chiefly, if not exclusively; as being more likely to be effective than ball.

3. But case-shot are unsuitable against a column, as they consist of a number of small balls which have not momentum enough to penetrate into it to any depth. When the enemy advances in columns, solid shot should plough through them from head to rear, a cross-fire being obtained upon them when possible.

4. Especially should round-shot be used against a close column rushing to attack us through a defile. Grape and canister might sweep down the leading ranks, but the mass of the column sees not the destructive effect; and being

carried forward by the pressure of the ranks behind, continues to rush on till the battery is carried, though with more or less loss.

At the battle of Seven Pines, on one occasion, when the rebels were advancing in close columns, they were plied with grape and canister from two of our batteries with but little effect. Every discharge made wide gaps in their leading ranks, but these were instantly filled again, and the columns pressed on.

A round-shot, on the other hand, comes shrieking and tearing its way through the entire column, carrying destruction to the very rear-most ranks. The hesitation produced is not limited to the leading ranks, but extends throughout the column. Thus both the actual and the moral effect of ball on a close column is much superior to that of grape or canister.

At the bridge of Lodi, the Austrian gunners plied the French column with grape. If they had used round-shot instead, it is doubtful whether Bonaparte would have succeeded in carrying the bridge.

5. But a direct fire with ball on the head of a column of attack would be much assisted by the fire of a few light pieces taking it in flank. And, for this fire in flank, case-shot would be powerfully effective; for, from their wide scattering, both their actual and their moral effects would extend through the whole column, from front to rear.

6. A battery with guns enough to keep up a continuous fire has little to fear from an open attack on it by infantry alone. Napoleon observed that no infantry, without artillery, can march one thousand yards on a battery of sixteen guns, well directed and served; for, before clearing two-thirds of the way, they will have been all killed, wounded, or dispersed.

This remark was made in reference to smooth-bore guns; the new rifled guns should be still more destructive.

7. Against skirmishers, as they offer but a poor mark for ball, grape and canister should alone be used.

E. AGAINST CAVALRY.

1. The ground round a battery should be so obstructed as to prevent the enemy's cavalry from closing on it; but in the case of a light battery, intended for manoeuvre, so far only as this can be done without impeding the movements of the battery itself.

2. On being threatened by cavalry, a light battery may sometimes promptly change its position to one where the cavalry would attack it at great disadvantage. For instance, if posted on an eminence, and cavalry should attempt to carry it by charging up the slope, instead of awaiting the charge in a position which would allow the cavalry to recover breath, and form on the height, it might run its pieces forward to the very brow of the slope, where the cavalry, having lost their impetus, and with their horses blown, would be nearly helpless and easily repulsed.

3. At a distance, the most effective fire on cavalry is with howitzers, on account of the terror and confusion caused among the horses by the bursting of the shells.

On the cavalry's flanks, when within four hundred or five hundred yards, the fire with grape would be most effective.

4. The last discharge should be of canister alone, and made by all the pieces at once, when the battery should be swiftly withdrawn.

5. If the cavalry succeed in reaching the pieces before their withdrawal, the gunners may find temporary refuge under the carriages.

F. AGAINST ARTILLERY.

1. As guns in the field should generally play on troops instead of batteries, there should be a reserve of horse artillery to take the enemy's batteries in flank, in case of need, as well as for other purposes.

2. Though the ordinary use of artillery is against infantry and cavalry only, special circumstances may render it expedient that a certain portion of it should respond to the enemy's batteries. In this case, not over one-third should be used for that purpose.

3. When we have guns in abundance, it might be sometimes advisable, by using several batteries at once, to silence the enemy's guns, before beginning to play on his troops.

4. Artillery properly plays on the enemy's guns--

(1.) At the moment of their coming into action, for then they are so exposed that our fire will be peculiarly effective, and perhaps to such a degree as to prevent their opening on us.

(2.) When our troops move forward to attack, in order to draw away from them the fire of the enemy's batteries, or, at least, to render it unsteady, inaccurate, and ineffective; and

(3.) Generally, when his guns are causing us very great damage.

5. Against guns, solid shot or shells only should be used, since they alone are capable of inflicting any serious injury on either guns or their carriages: solid shot, by their great weight and momentum; shells, by their bursting.

But within three hundred or four hundred yards, grape and canister would soon destroy the gunners and the horses.

6. Our fire on a hostile battery would not only be very effective when it is coming into action, but at all other times when its flanks are exposed; as in

limbering up to move off, or in a flank march. On these occasions we should use grape, if near enough; otherwise, spherical case would be generally the best.

7. When possible to avoid it, a field battery should not be opposed to a battery of position, or, generally, a battery of light guns to one of heavy guns. For even when the numbers of pieces on the two sides are equal, the enemy's superiority in range and in weight of metal would give him such advantage in the duel that our own battery would soon be destroyed or silenced.

VI.--Its Fire.

1. It is important not to commence the fire till our guns are in sure range--

(1.) Because a harmless fire serves but to embolden the enemy and discourage our own troops.

(2.) Because artillery ammunition should never be wasted. The fate of a battle will sometimes depend on there being a sufficient supply of it at a particular moment.

2. The usual maximum distances at which smooth-bore field guns may open fire with any considerable effect, are--

For 12-pounders 1100 yards. For 6 " 750 "

What these distances are in respect to rifled guns, it does not appear to be yet definitively settled. The extreme range of the new rifled six-pounders is said to be three thousand yards; of the twelve-pounders, four thousand five hundred yards.

3. Guns are usually fired by order, and not by salvos, or volleys; and never as soon as loaded, unless delay be dangerous. Artillery fire is formidable only in

proportion to its accuracy; and this is attainable only by a cool and calculated aim.

4. Artillery fire should never cease through the whole line at once. This would have a discouraging effect on our own troops, and an inspiriting one on the enemy's. Especially must this not be done when we are about to execute any manoeuvre; for it would be sure to call the enemy's attention to it.

Therefore, if a particular battery has to change its position, it must not cease the fire of all its pieces at once; for, besides its disquieting effect on our troops, it would notify the enemy of the movement.

5. In covering an attack, our guns should keep up their fire till the moment it would begin to endanger our advancing columns.

6. By a ricochet fire, artillery is said to increase its effect, from one-fourth to one-half. It is especially effective in enfilading a line of troops, a battery, or the face of a work taken in flank.

Ricochet shots have also great moral effect.

7. When used against breastworks of rails or logs of wood, guns should be fired with moderate or shattering charges; so as more surely to demolish them, and, at the same time, to increase the destructive effect of the fire by scattering the splinters.

In view of the frequent necessity of battering such defences, and of using a ricochet fire, which also requires small charges of powder, it would be an improvement in our artillery service to make a certain proportion of the ammunition in each gun limber to consist of cartridges of half the usual size.

8. In bombarding a village during a battle, if our object be to set fire to and destroy it, this will be best accomplished with shells. If we wish merely to

drive the enemy from it, solid shot from heavy calibres will be necessary, which will more surely reach and destroy his troops, wherever they may show themselves.

9. The moral effect, both of solid shot and of shells, is much greater than that of grape or other case-shot, from their more fearful effects on the human frame, and from the great number of men that a single ball or shell will kill or fatally wound. One twelve-pound solid shot has been known to kill forty-two men, who happened to be exactly in its range.

10. Ball and shell should be used--

(1.) When the enemy is at a distance. (2.) When he is in mass. (3.) When he is in several lines. (4.) When his line may be enfiladed.

11. In silencing a battery, our fire should be concentrated on one piece at a time, while some of our guns throw spherical case, from a raking position, if possible, on the gunners.

12. Though grape has a much less range than ball, yet within four hundred yards, on account of its scattering, its effect is superior to it.

The fire of guns double shotted with ball and a stand of grape, is fearfully destructive when used at a short distance to repel a charge.

13. It is artillerymen's point of honor not to abandon their pieces till the last extremity. They should always remember that when the enemy is closing on them, the last discharge will be the most destructive of all, and may suffice to repulse the attack and turn the tide of battle.

In our attack on Quebec, in our War of Independence, on the near approach of our assaulting column, the British gunners fled from their pieces; but the one who fled last, before leaving, discharged his gun, which was loaded with grape. The discharge killed and wounded several of our men; among them the brave General Montgomery, who was leading the column in person. The effect was decisive. The assailants fell back, panic-struck at the loss of their commander; and our attack failed.

14. To prevent our own captured pieces from being turned on us by the enemy, it is the artillerymen's last duty, when it has become evident that the guns cannot be saved, to spike them. The operation requiring but a few moments, it can rarely be omitted without disgrace.

VII.--Its Supports.

1. Artillery must always be protected from the melee, in which it is helpless whether against infantry or cavalry; and should therefore have a strong support.

2. A portion, at least, of every artillery support, should consist of sharpshooters, whose fire will be the most effective in driving off skirmishers sent forward to threaten or attack the guns, or to pick off the gunners.

3. Batteries must be secured on both flanks, and their supports will be posted with that view; on one or on both flanks, according to circumstances, and more or less retired.

4. A support should never be directly in rear of the battery, where it would be in danger, not only of receiving the shots aimed at the battery, but also of killing the gunners belonging to it, as has repeatedly happened during the late war.

5. A battery and its support owe each other mutual protection. Therefore, when an infantry support, after holding the enemy in check, to enable the battery to limber up and retire, is obliged, in its turn, to retreat, the battery should take such a position as will enable it to cover the retreat by its fire.

6. A battery consisting of any considerable number of pieces may be able,

sometimes, to dispense with a support.

During the battle of Ucla, in Spain, in 1809, the French artillery, under General Senarmont, had been left behind, owing to the badness of the roads. The Spaniards came down upon it in large force. On their approach, the guns were formed in square, and, by their fire on every side, defended themselves successfully, and finally repulsed the enemy.

So, at Leipsic, in 1813, when the grand battery of the Guard, which had been left temporarily without a support, was charged by the Russian hussars and Cossacks, Drouot, its commander, rapidly drew back its flanks till they nearly met; and the cavalry were soon repulsed by its fire.

7. At the battle of Hanau, in 1813, Drouot's eighty guns were charged by the Allied cavalry. Having no supports, he placed his gunners in front of them with their carbines. This checked the cavalry, who were then covered with grape, which drove them back to their lines.

Ought not our gunners also to have carbines slung on their backs for such emergencies? Against infantry, our guns would often need no other support. The carbines could be used to reply to the enemy's skirmishers; and the example just cited shows that, thus equipped, artillerymen may sometimes successfully defend their pieces even against cavalry.

TACTICAL USE OF CAVALRY.

We shall consider--

I.--ITS FORMATIONS. II.--ITS STRONG AND ITS WEAK POINTS. III.--HOW IT IS POSTED. IV.--ITS SUPPORTS. V.--HOW IT IS USED. VI.--HOW IT FIGHTS. VII.--ITS CHARGE. VIII.--ITS ATTACK ON INFANTRY.

And shall conclude with some--

IX.--GENERAL REMARKS.

I.--Its Formations.

1. The formations of cavalry for battle are either--

(1.) In deployed lines. (2.) Lines of regiments, in columns of attack doubled on the centre. (3.) A mixed formation of lines and columns. (4.) Echelons of lines or columns; or-- (5.) Deep columns.

2. Deployed lines are not objectionable in principle. They are often not advisable; but are sometimes necessary.

But long, continuous lines should always be avoided; being unfavorable to rapid manoeuvring, which cavalry is constantly called upon to execute in the field.

3. Cavalry has, in its horses, an unreasoning element, which is not controllable, like men; and is therefore much more easily thrown into disorder than infantry. For this reason, when deployed, it should always be in two lines; the second behind the first; the first line deployed, and the second in columns of squadrons by platoons. There should be also a reserve at a few hundred paces behind the second line.

The second line should be near enough to the first to be able to support it, if checked; but not so near as to partake in its disorder, if repulsed.

4. Cavalry should be always in column when expecting to manoeuvre, or to be called on to make any rapid movement; this being the formation best adapted for celerity.

5. Cavalry deploys in lines--

(1.) When preparing for a charge in line.

(2.) When preparing for any attack requiring the utmost possible width of front; as where the enemy is to be suddenly surrounded.

(3.) When it becomes necessary in order to prevent our troops from being outflanked by the enemy's.

(4.) When exposed to continued artillery fire, which is much less destructive on a line than on a column.

6. Cavalry should always present a front at least equal to the enemy's; otherwise, its flank, which is cavalry's weak point, will be exposed to attack. When inferior in numbers, we may extend our line by leaving intervals, more or less wide, between its different corps. Any hostile squadrons that may attempt to pass through these intervals to take the line in rear, could be taken care of by the second line.

7. The best formation in respect to mobility is a line of regiments in columns by squadrons, doubled on the centre; corresponding to infantry double columns.

8. The mixed formation of lines and columns is more manageable than simple lines. Which of these two is preferable depends upon the ground, and upon all the other circumstances of the case.

9. The order in echelons is as good in attack as in retreat; since the echelons mutually support each other.

10. Decidedly the most objectionable of all cavalry formations is that in deep columns:

(1.) From the almost entire loss which it involves of its sabres, which are cavalry's peculiar and most effective weapon.

(2.) From the long flanks which it exposes to attack.

11. The formation in one rank, instead of two, has been introduced by the new Cavalry Tactics, though it has been as yet but partially adopted in the field.

This innovation has two advantages. It doubles the number of sabres to be used against the enemy; and it enables the cavalry to cover double the ground; thus doubling, also, its power to outflank, which is a valuable advantage, especially when opposed to cavalry.

Its disadvantage is, that it must, more or less seriously, impair the solidity and vigor of the cavalry charge proper; in which a whole line, with "boot to boot" compactness, comes at once to the shock, like some terrific mechanical engine; and in which the riders in the front rank are compelled to dash on with full speed to the last; knowing that if they slacken rein, even for a moment, they would be ridden over by the rear-rank men one yard behind them. From there being no rear-rank to fill up the gaps caused, during the charge, by the enemy's missiles, or by casualties occasioned by obstacles of the ground, the charging line must generally arrive on the enemy broken and disunited, or as foragers. The moral effect of such a charge on our own men will be unfavorable, as they will not realize the certainty of mutual support at the critical moment; and its moral effect on the enemy must be decidedly inferior to that produced by a charge that is at once swift, solid, and compact.

But the force of this objection is somewhat weakened, by the consideration that the compact charge of "cavalry of the line" must hereafter be comparatively rare, in consequence of the introduction of rifled artillery and infantry weapons, with their greatly increased accuracy and range; which ought to cause such slaughter in a line or column of charging cavalry, that, if it arrive at all to the shock, it would generally be only in scattered groups.

12. In advancing over wooded, or other obstructed ground, it may be necessary to break the line into company columns of fours, as in the infantry

manoeuvre of advancing by the flanks of companies.

As the cavalry column of fours corresponds to the march of infantry by the flank, the use of this formation in action is open to the same objections that have been already pointed out as applying to flank marches by infantry.

II.--Its Strong and its Weak Points.

1. The value of cavalry on the battle-field consists chiefly in its velocity and mobility. Its strength is in the sabre-point and spurs.

2. Its charge is accompanied with a powerful moral effect, especially upon inexperienced troops. But,

3. Cavalry has but little solidity, and cannot defend a position against good infantry. For, if it remain passive on the ground it is to hold, the infantry will soon destroy it by its fire, to which it cannot, with any effect, reply; and if it attack at close quarters, the infantry, by means of its defensive formations, will be able, at least, to hold its ground, and probably repulse its charges by a reserved fire. So that the cavalry will finally have no alternative but to retire.

4. It is exposed and helpless during a change of formation; like artillery limbering up, or coming into action.

5. On its flanks, it is the weakest of all arms. A single squadron attacking it suddenly in flank, will break and rout cavalry of ten times its number.

At the battle of the Pyramids, Napoleon kept a few squadrons in rear of either flank, which, on his line being charged by a formidable body of Mamelukes, vastly superior to his own cavalry in numbers, horses, and equipments, nevertheless suddenly fell on their flanks and destroyed them.

6. Cavalry is never so weak as directly after a successful charge; being then exhausted, and in more or less disorder.

III.--How Posted.

1. A part of our cavalry must be so posted as to secure our flanks; remaining in column behind the wings, till the enemy's movements require its deployment.

If one wing is covered by natural obstacles, give the cavalry to the uncovered wing; posting it in rear of the flank battalion of the second line.

2. When cavalry is posted on the flanks, it should not usually be on the first line of infantry. If it is to be used for attack, it is better to keep it retired from view till the last moment, in order to strengthen its attack by the powerful moral effect of a surprise. And, used defensively, it will be best posted on the flanks of the second line; since, in advancing to charge, it must have a clear space in its front of at least two hundred or three hundred yards, to enable it to act with freedom and vigor.

3. But if a position can be found for cavalry in front, where it would not be too much exposed, this may sometimes enable it to exercise an important moral effect, by threatening the flank of such of the enemy's troops as may be sent forward to attack.

At the battle of Leipsic, in 1813, the Wurtemburg cavalry was launched against Blucher's Prussian cavalry. But, seeing the Prussians drawn up not only in front, but opposite their flank, they lost confidence, charged feebly, and too late. They were consequently repulsed and driven back on the Marine Battalion, which they threw into confusion.

So, at the battle of Prairie Grove, in December, 1862, the First Iowa Cavalry, which was held in reserve, by its mere presence, caused every attempt of the rebels' flanking regiments to be abandoned.

4. In order not to impede the manoeuvres of the infantry, cavalry should not

fill intervals in the lines, or be placed between the lines.

It is dangerous when the ground is such as to require the cavalry in the centre of the first line; for, if it is beaten, a gap is left through which the enemy may penetrate. At the battle of Blenheim, in 1704, Marlborough owed his victory, in great measure, to the Allies' forcing back the cavalry forming the centre of the French army; thus turning the whole of its right wing, and compelling the infantry posted at Blenheim to surrender.

5. Yet cavalry should always be near enough to the infantry to take immediate part in the combat; and although it should not be posted in the intervals between infantry corps, it may debouch through them, in order to attack more promptly.

At the battle of Friedland, the Russian cavalry charged a French infantry division. Latour Maubourg's dragoons and the Dutch cuirassiers, riding through the battalion intervals, charged the Russians in turn, and drove them back on their infantry, throwing many of them into the river.

6. When both wings are uncovered, the best place for the cavalry will usually be in rear of the centre of the second line; whence it can be sent in the shortest time to either wing.

7. Cavalry should not be scattered over the field in small detachments, but be kept massed at one or more suitable points; as behind the centre, or behind one wing, or both wings. A small cavalry force should be kept entire; or it will have very little chance of effecting any thing whatever.

Cavalry of the line, to produce its decisive effects, must be used in heavy masses. In the beginning of the Napoleonic wars, the French cavalry was distributed among the divisions. Napoleon's subsequent experience led him to give it more concentration, by uniting in one mass all the cavalry belonging to each army corps; and, finally, these masses were again concentrated into independent cavalry corps; leaving to each army corps only cavalry enough to guard it.

8. For tactical operations in the field, cavalry insufficient in number is scarcely better than none at all, as it can never show itself in presence of the enemy's cavalry, which would immediately outflank and destroy it, and must keep close behind its infantry.

At the opening of Napoleon's campaign of 1813, he had but very little cavalry to oppose to the overwhelming masses of this arm possessed by the Allies. In consequence of this, he could make no use of it whatever; and the tactical results of the battles of Lutzen and Bautzen were far inferior to those habitually obtained in his former victories, and were purchased with much greater loss.

9. Small bodies of cavalry threatened by the enemy's cavalry in greatly superior force, may sometimes be saved by taking refuge in an infantry square, as practised by Napoleon at the battle of the Pyramids.

10. Cavalry should remain masked as long as possible; for it produces most effect when its position and movements are hidden, so that a strong force may suddenly be brought upon a weak point.

For this reason, a flat, open country is less favorable for this arm than plains with undulations, hills with gentle slopes, woods, villages, and farms; all these being so many facilities for screening cavalry from view.

11. Cavalry should never be brought to the front, except to engage. It is unfortunate when the ground is such as to prevent this; for cavalry, compelled to remain inactive under fire, is in great danger of becoming demoralized.

12. As to the ground:

(1.) Cavalry must not rest its flank on a wood, a village, or other cover for an enemy, till it has been occupied by our own troops. If compelled to do so, it should send out patrols to reconnoitre and observe. Its position is no longer tenable from the moment the enemy appears within striking distance on its flank.

(2.) It must not be posted on the very ground it is to defend, but in rear of it; as it acts effectively only by its charge.

Attacking cavalry must have favorable ground in front; defending cavalry, in rear. An obstacle in either case may be fatal.

IV.--Its Supports.

1. The flanks of cavalry lines or columns are always exposed. They should, therefore, be protected by supports of light cavalry, which can act promptly and swiftly. When behind a line, these supports should be usually in open column, so as to be able to wheel, without a moment's delay, into line.

2. The most effectual mode of protecting the flank of a line or column of cavalry is by means of squadrons in rear, formed in echelons extending outwards; as this exposes the enemy's cavalry that may attempt to charge the main body in flank to be immediately charged in flank themselves; which would be destruction. For this purpose, irregular cavalry may be as effective as any other.

3. This cavalry support or reserve behind the flanks may sometimes play an important offensive part. The enemy's first line, the instant after either making or receiving a charge, is always in greater or less disorder; and a vigorous charge then made on it in flank by our own flank reserve, would have a decisive effect.

4. Cavalry should never engage without a support or reserve in rear, not only to guard its flanks, but also to support it when disorganized by a successful charge.

5. So, when engaged in skirmishing order, being then very much exposed, it must always be protected, like infantry skirmishers, by supports in close order.

6. It has been already seen that, although cavalry may carry a position, it cannot hold it, if attacked by infantry. When used for such a purpose, therefore, it should always be accompanied by an infantry support.

The French cavalry succeeded in carrying the plateau of Quatre Bras; but, having no infantry with it to reply to the terrible fire of the Allied infantry from the surrounding houses, it was compelled to retire, and yield it again to the enemy.

According to Wellington, Napoleon frequently used his cavalry in seizing positions, which were then immediately occupied by infantry or artillery.

V.--How Used.

1. Cavalry generally manoeuvres at a trot. At a gallop, disorder is apt to take place, and exhaustion of strength that will be needed in the charge.

2. The ordinary use of cavalry is to follow up infantry attacks and complete their success. It should never be sent against fresh infantry; and should generally, therefore, be reserved until towards the last of the action.

Napoleon, who, by concentrating his cavalry into considerable masses, had enabled himself to use it on the battle-field as a principal arm, sometimes produced great effects by heavy cavalry charges at the very beginning of the action.

But, though Napoleon's splendidly trained heavy cavalry might sometimes break a well-disciplined infantry without any preparatory artillery fire, it would be dangerous to attempt this with cavalry inferior to it in solidity; and the new rifled weapons would seem to render the cavalry charges of his day no longer practicable.

3. Cavalry may be hurled against the enemy's infantry--

(1.) When it has been a long time engaged, and therefore exhausted.

(2.) When it has been shattered by artillery.

And always should be--

(1.) When it is manoeuvring.

(2.) When the attack would be a surprise.

(3.) When its ranks begin to waver, or when it manifests any unequivocal symptom of hesitation or intimidation.

In the three latter cases, success will usually be certain; in the two former ones, quite probable: but, in most other cases, a cavalry charge will succeed, perhaps, only one time in ten.

4. The chief duties of cavalry in a defensive battle are--

(1.) To watch the enemy's cavalry, to prevent its surprising our infantry.

(2.) To guard our troops from being outflanked.

(3.) To defend our infantry and artillery while manoeuvring.

(4.) To be ready to charge the enemy the instant his attack on our troops is repulsed.

5. Used offensively, it must promptly attack--

(1.) The enemy's flanks, if uncovered.

(2.) His infantry, when, from any cause, its attack would probably succeed.

(3.) All detachments thrown forward without support.

6. When cavalry has routed cavalry, the victorious squadrons should at once charge in flank the infantry protected by the cavalry just beaten. The great Cond? when only twenty-two years of age, by this means, won the victory of Rocroi.

7. Deployed as skirmishers, by their noise, dust, and smoke, cavalry may furnish a good screen for our movements.

8. Cavalry skirmishers scout their corps, to prevent the enemy reconnoitring it too closely.

9. When a cavalry rear-guard has to defend, temporarily, a defile, a bridge, or a barricade, a part should dismount, and use their carbines till the rest are safe.

So, a cavalry vanguard, by its fire, dismounted, may prevent the enemy from destroying a bridge.

In these, and in similar cases, the cavalrymen should habitually dismount, in order to render their fire effective; acting and manoeuvring as skirmishers.

VI.--How it Fights.

1. The success of cavalry in battle depends on the impetuosity of its charge, and its use of the sabre. When deployed as skirmishers, mounted or dismounted, its proper weapon is the carbine or pistol; and in individual combats, these weapons may occasionally be very useful. But when acting as cavalry proper, in any compact formation, it must rely on the sabre. The aim

with a pistol or carbine in the hands of a mounted man is so unsteady, that the fire of a line of cavalry is generally ineffective; and there are few occasions where it should be resorted to. When cavalry has learned to realize that these are not its true arms, and that it is never really formidable but when it closes with the enemy at full speed and with uplifted sabre, it has acquired the most important element of its efficiency.

2. Cavalry should, therefore, not fight in columns, as most of its sabres would thereby become useless. But if a facing about to retreat is feared, an attack in column would prevent it. It is said, also, that a column is more imposing than a line. If so, it might have a greater moral effect on the enemy.

3. When cavalry are deployed as skirmishers, as a curtain to hide our movements, they should be in considerable number, with small intervals, and should make as much noise, and smoke, and dust as possible. When the charge is sounded, the skirmishers wait and fall in with the rest.

4. The great rule in cavalry combats is to cover our own flanks, and gain the enemy's; for these are his and our weakest points.

5. When the enemy's cavalry is already in full charge on our infantry, it is too late for our cavalry to charge it with much prospect of success. In such a case, it would be better to defer our own charge till the moment that the enemy's is completed; for our success then would be certain.

6. Cavalry attacks cavalry in line, in order to have the more sabres, and, if possible, to outflank the enemy.

7. If we can manoeuvre so as to attack the enemy's cavalry in flank, our success will be certain.

Military history affords hundreds of instances in proof of this proposition. At one of the battles in Spain, for example, in 1809, fifteen hundred French horse, by charging four thousand Spanish cavalry in flank, completely cut it in pieces.

8. Cavalry never waits in position to be charged by cavalry. Its only safety is in meeting the charge with a violent gallop; it would otherwise be sure to be overthrown.

When hostile cavalries thus meet each other, there is usually but small loss on either side. A certain number of troopers are usually dismounted; but the colliding masses somehow ride through each other, allowing but little time for the exchange of points and cuts.

Thus cavalry can defend itself against cavalry only by attacking; which it must do even when inferior to the enemy in number.

9. To attack artillery, cavalry should be in three detachments; one-fourth to seize the guns; one-half to charge the supports; and the other fourth as a reserve.

The first party attacks in dispersed order, as foragers, trying to gain the flanks of the battery. The second party should manoeuvre to gain the flanks of the supports.

10. Where a cavalry attack can be masked, so as to operate as a surprise, a battery may be taken by charging it in front. The formidable Spanish battery in the Pass of Somosierra, was finally carried by a dash of Napoleon's Polish Lancers upon it, suddenly profiting of a temporary fog or mist. But, in ordinary cases, when cavalry has to charge a battery in front, its fire should be drawn by our own guns or infantry, immediately before the charge begins.

11. In an attack on an intrenchment, the office of cavalry can rarely be any thing else than to repulse sorties from the work, and to cut off the enemy's retreat from it.

VII.--Its Charge.

1. As cavalry acts effectively on the field of battle only by its charge, good cavalry of the line can be formed in no other way than by being exercised in this, its special and peculiar function.

On taking command of the Army of Italy in 1796, Bonaparte found the French cavalry to be entirely worthless. They had never been accustomed to charge, and he had the greatest difficulty in making them engage. Seeing the great importance of this arm, he determined to make good cavalry of them by compelling them to fight. So, in his attack on Borghetto, he sent his cavalry forward, with his grenadiers on their flanks, and his artillery close behind them. Thus enclosed, and led on by Murat to the charge, they attacked and routed that famous Austrian cavalry whose superiority they had so much dreaded. This was the first step in the formation of the splendid French cavalry to which Napoleon afterwards owed so many of his victories. And, at the battle of Hochstedt, on the Danube, in 1801, its superiority over the Austrian cavalry was, at last, completely established.

2. Cavalry charges--

(1.) In line; but this only on even ground, and at short distances;

(2.) In column; and

(3.) As foragers, or in dispersed order. But this kind of charge is exceptional. It can rarely be used with safety against any but an uncivilized or an undisciplined foe.

3. A charge in one long continuous line should never be attempted. Such a charge will be usually indecisive, as it cannot be made with the necessary ensemble or unity. The success of a charge in line depends on the preservation of a well-regulated speed and of a perfect alignment; by means of which the whole line reaches the enemy at once. At the charging gait, this is rarely attainable; so that the charge in line, except at short distances, and

over very even ground, usually degenerates into a charge by groups, or individual troopers, arriving successively. The most dashing riders, or those mounted on the swiftest horses, will naturally arrive first, and be overpowered by numbers.

4. A charge in deep column is also objectionable; its long flank exposing it too much to artillery fire and to the enemy's cavalry.

But when cavalry is surprised, it must charge at once, in whatever order it happens to be, rather than hesitate or attempt to manoeuvre, for this would expose it to destruction.

5. A prompt and unhesitating obedience to the command to charge, without regard to the circumstances under which it is given, may sometimes lead to results unexpected even to the charging troops themselves.

One instance of this was related to me by an old officer of Napoleon's favorite Fifth Cuirassiers. The regiment was on the left of the line of battle. Directly in front of it was an extensive marsh; beyond which rose an eminence, abrupt in front, but sloping gently towards the rear, the crest of which was crowned by formidable Austrian batteries. For two hours the cuirassiers had been standing in line, listening to the roar of battle on the right, and eagerly expecting a summons to go somewhere to engage the enemy. The very horses were neighing and pawing the ground, in their impatience to be off. Just then galloped up one of the Emperor's aids, saying, "Colonel, the Emperor desires you to charge directly on the enemy's batteries opposite your position." The brave colonel, who was one of Napoleon's personal favorites, though chafing at the prolonged inaction of his command, pointed to the marsh, and requested the officer to inform the Emperor of the obstacle in his front, with the existence of which his majesty, he said, was probably unacquainted. In a few minutes the officer came riding furiously back with a message to the colonel, that "if he did not immediately charge, the Emperor would come and lead the regiment himself." Stung by this reproof, the colonel plunged his spurs into his horse's flanks, and giving the command "Forward," led his regiment, at full gallop, directly through the marsh upon the point that had been indicated.

The charge itself was, of course, a failure. The regiment finally struggled its way through the marsh to the opposite side, but leaving behind it a large number of gallant officers and men, who had sunk to rise no more; my informant being of the number who escaped.

But the result of this demonstration was most decisive. Seeing that the height on which the Austrians had planted their heavy batteries, and which commanded the entire ground, was the key of the battle, Napoleon had determined to wrest it from them, together with the batteries which crowned it. Accordingly, the evening before, he had dispatched a body of light infantry by a very circuitous route, to turn the position and attack the batteries in rear. He had accurately calculated the time the detachment would require to reach its destination; and when the moment arrived at which it should be ready to commence its attack, he ordered the cuirassiers to charge directly upon the position in front. The Austrian artillery, suddenly attacked in rear, and, at the same time, threatened with a cavalry charge in front, where it had deemed itself perfectly secure, tried to change the position of its pieces, so as to get a fire on its assailants from both directions. But it was too late; the temporary confusion into which it was thrown enabled the French infantry to carry all before it, and the height was won, with all its batteries.

So, at Waterloo, Sir Hussey Vivian's brigade of Light Cavalry, which was marching in column by half squadrons, left in front, had begun to form up into line on the leading half squadron, when an order arrived from Wellington to charge. Instantly the charge was made, and, of course, in echelons of half squadrons, extending to the right. The effect of this was that a body of French cavalry on its right, then attacking the British line, was suddenly taken in flank and completely routed.

6. A charge in deep column may sometimes be made necessary by the

nature of the ground, which, at the same time, protects its long flanks: as where, in our pursuit of the rebels after the battle of Nashville, in 1864, the Fourth United States Cavalry, approaching them over a narrow turnpike, made a vigorous charge in column of fours, which broke their centre, and, with the help of infantry skirmishers on the flanks, drove them from the ground.

7. When the ground is rugged, in order to lessen the number of falls, the rear-rank, in the charge, should open out six paces, closing up again at the last moment.

8. Cavalry advances to charge at a trot, or at a gallop. A fast trot is better than a gallop, as alignments are not easily kept at great speed. Experience has shown that the best distance from the enemy to begin the gallop, is about two hundred and sixty yards; thence steadily increasing to the maximum of speed. This gradual increase of speed is very important, to prevent the horses from being completely blown on reaching the enemy.

9. Cavalry should not charge by a wood, till it has been carried by our own infantry, if it can possibly be avoided.

At the battle of Kollin, in 1756, Frederick's cavalry, pursuing the Austrians, was taken in flank by some Austrian infantry posted in a wood, and made to retire with great loss.

10. When cavalry is required to charge over unknown ground, it should be preceded by a few men thrown out to the front as skirmishers, in order to scout the ground to be passed over. The neglect of this precaution has sometimes led to great disaster.

At Talavera, two cavalry regiments, the First German Hussars and the Twenty-third Light Dragoons, were ordered to charge the head of some French infantry columns. When near the top of their speed they came suddenly upon a deep ravine, with steep sides. Colonel Arentschild commanding the Hussars, who was in front, at once reined up, and halted his regiment, saying: "I vill not kill my young mensch!" But the other regiment, commanded by Colonel Seymour, which was on its left, not seeing the obstacle in time, plunged down it, men and horses rolling over on each other in frightful confusion. Of the survivors, who arrived on the other side by twos and threes, many were killed or taken; and only one-half of the regiment ever returned.

So, at the battle of Courtrai, in 1302, from the French cavalry's omitting to scout the ground they charged over, the Flemings won a great victory. All the 閘 ite of the French nobility and chivalry was destroyed, and gold spurs were collected by bushels on the field. It was the French Cann? The Flemings were drawn up behind a canal, flowing between high banks, and hidden from view. The French rushing on at full gallop, all the leading ranks were plunged into the canal. The entire cavalry was thereby checked and thrown into irretrievable disorder, which extended to the infantry, in their rear. The Flemings, profiting by their confusion, crossed the canal at two points simultaneously, attacked them in flank, and completed their rout.

So, at the battle of Leipsic, in 1813, Murat, in his great cavalry charge on the Allied centre, had captured twenty-six guns, and was carrying all before him, when he pushed on to the village of Gulden Gossa, where the ground had not been reconnoitred, and could not be distinctly seen from a distance. Here the French found their career suddenly checked by a great hollow, full of buildings, pools of water, and clusters of trees; while the Allied infantry, from behind the various covers afforded by the ground, opened upon them a destructive fire. Being then suddenly charged in flank by the Russian cavalry, they were driven back with heavy loss; the Allies recapturing twenty of the twenty-six guns they had lost.

The troopers employed to scout the ground before a charge would not be in much danger from the enemy, who would hardly fire on a horseman or two, especially when expecting a charge.

11. Cavalry must never pursue, unless its supports are close at hand.

In pursuing, it must be circumspect, and not go too far. Union and order are indispensable; for, without them, a slight resistance may suffice to cause a repulse.

VIII.--Its Attack on Infantry.

First, as to its ATTACK GENERALLY. Secondly, its attack ON SQUARES.

A. GENERALLY.

1. Cavalry must avoid distant engagements with infantry; in which the latter must always have the advantage.

2. The slightest cavalry charge on the flank of infantry will rout it.

3. As to a cavalry attack in front: If the infantry stand firm, the chances are against its success. If the infantry cannot be attacked in flank, the cavalry should therefore wait till it has been shattered by artillery, or has become exhausted, or demoralized, or till it begins to manoeuvre.

4. If the infantry be in line, or in column, cavalry should attack it in flank; if in square, at one of its angles; if in several squares, at one of its flank ones, so as to avoid a cross-fire from the other squares. If a flank square be broken, the next one to it, being no longer protected by the fire of any other square, may be attacked with the same prospect of success; and so on successively.

5. But if the hostile infantry have supporting cavalry, we must not charge in such a manner as to enable it to take us in flank.

6. To test the infantry about to be attacked, cavalry may pass a few hundred paces in its front, to threaten it, sending forward a few horsemen to fire, gallop forward, and raise a dust. If the infantry, instead of disregarding these

movements, begin to fire, it will probably be broken, on the cavalry's charging it at once and vigorously, whether in column or in line. But otherwise, if the infantry reserve its fire, and only sends out a few sharpshooters.

7. Ascending slopes, if not too steep, are not unfavorable to attacks on infantry; for their shots, as experience shows, will then mostly fly too high.

8. On a descending slope, cavalry charges down on infantry with terrible effect; as it then arrives with an impetus which nothing can stop.

At Waterloo, a column of French infantry was ascending a steep slope. Suddenly the Scotch Greys cavalry regiment dashed down upon it from above, rode over, and destroyed it.

B. ON SQUARES.

1. Infantry squares are usually charged in open column; the distance between the subdivisions being a subdivision front and a few yards over; in order that each subdivision may have time to break into the square, or, if unsuccessful, to disengage itself and retire.

But the distances should never be so great as to allow the square to reload after firing a volley at the next preceding subdivision.

2. The leading subdivision will usually draw the fire of the square. If this is delivered at very short range, say at twenty paces, it will raise up a rampart of dead and wounded men and horses which will probably suffice to check the following subdivisions, and so repulse the charge. But an infantry square rarely reserves its fire so long; and if the fire is delivered at any considerable distance, no such effect will be produced.

3. A good formation to attack a square is said to be a column of three squadrons, with squadron front, at double distance; followed by a fourth squadron, in column of divisions or platoons, to surround the square, and

make prisoners, if it is broken.

4. Before cavalry charges a square, it should be first shattered or demoralized by artillery fire, when this is practicable. In the absence of artillery, sharpshooting infantry skirmishers may, to a certain extent, supply its place.

5. A square should be attacked at one of its angles, which are obviously its weakest and most vulnerable points. But to cover a real attack on an angle, cavalry sometimes makes a false attack on the front of a square.

6. When squares are formed checkerwise, cavalry must attack a flank square, and not expose itself to a cross-fire by charging an interior one.

7. Cavalry charging a square firing irregularly will usually break it. But when the square reserves its fire, and pours in well-aimed volleys at short range, the charge will rarely succeed. The cavalry should, therefore, before charging, use every effort to draw the fire of the square, or of the fronts which threaten it. This is sometimes accomplished by sending forward a few skirmishers to fire on the square.

8. When one square fires to assist another, the firing square should be instantly charged, before it has time to reload.

9. To succeed, a cavalry charge should be made with a desperate, forlorn-hope recklessness, and with reiterated attacks on one point. If the fire has been delivered at very close range, though its effect has probably been destructive, the smoke will momentarily shut out the line of infantry from the horses' view, thus removing the chief obstacle to their breaking through it. The survivors of the fire should therefore rush desperately on.

If the French attacks on the British squares at Quatre Bras had been made in this manner, instead of opening to the right and left, and diverging to a flank at the moment of closing, they would probably have succeeded.

But this sudden divergence is often the fault of the horses, which instinctively recoil before a serried line of infantry, with bayonets at the charge. Cavalry should, therefore, never be practised on the drill-ground in charging a square, as the horses would thereby acquire the habit of suddenly checking their course, or of diverging to a flank, on arriving at the enemy. This would so strengthen their natural instinct that they could never be got to break a square. Or, at least, when this manoeuvre is practised for the purpose of instruction, the horses used should never afterwards be taken into the field.

10. The cavalry most formidable to an infantry square are Lancers. Their lances, which are from eleven to sixteen feet long, easily reach and transfix the infantry soldier, while the sabres of the other cavalry are too short to reach him over the horse's neck, and over the musket, lengthened by the bayonet. But Lancers are usually no match against other cavalry, who can parry and ripost before the lance can resume the guard.

11. When cavalry has succeeded in completely breaking a body of infantry, it may often inflict fearful slaughter upon them.

At the battle of Rio Seco, in Spain, after Lasalle's twelve hundred horse had broken the Spanish infantry, they galloped at will among twenty-five thousand soldiers, some five thousand of whom they slew.

IX.--General Remarks.

1. Besides its uses on the field of battle, cavalry may render most important service in completing the destruction of beaten corps, or compelling their surrender, and so enable us to secure the great strategic objects of the campaign. Thus, after the battle of Waterloo, it was the Prussian cavalry that completed the dispersion of the French army, and prevented it from rallying. And, but for Napoleon's ill fortune in respect to Grouchy, in that battle, he would, to all appearance, have succeeded in accomplishing his plan of

campaign, which was, to separate the English from the Prussians, beat them in detail, and complete their destruction with his twenty thousand cavalry.

2. The battles of the late War of the Rebellion, the earlier ones, at least, were mostly indecisive. One chief cause of this was, that neither side had a sufficient force of true cavalry to enable it to complete a victory, to turn a defeat into a rout, and drive the enemy effectually from the field. The cavalry charges were generally such as mounted infantry could have just as well made; charges in which the pistol and carbine played the principal part, instead of the spur and sabre. It was not until the fight at Brandy Station, in June, 1863, that sabres were used, to any extent, at close quarters. Thus, neither of the contending armies was able to break up and disperse, destroy, or capture its enemy's infantry masses, in the manner practised in Napoleon's great wars, not having, to any considerable extent, that description of force called Cavalry of the Line, which alone is capable of effecting these results by its solid and compact formations, its skilful, yet rapid manoeuvring, and its crashing charges.

3. European cavalry of the line is divided into Heavy and Light. Heavy cavalry is heavily armed; that is, their weapons are larger and heavier than those of light cavalry, and to these weapons, carbines, in most of the corps, are added. Some of the corps wear steel or brass cuirasses; and the men and horses are of the largest size.

In Light cavalry, the only weapons are the sabre and pistol; and the men and horses are light and active, rather than strong and large.

Lancers are considered a medium between Heavy and Light cavalry.

4. Great as may be the advantages of a large force of regular cavalry of the line, there were serious objections to its being raised at the opening of the late war.

(1.) The theatre of war presented nowhere any of those wide and level plains so common in Europe, and on which cavalry masses are able to produce such decisive effects in battle. On the contrary, the ground was almost everywhere so rugged and mountainous, or else so densely wooded, as to be extremely unfavorable to the movements of cavalry of this description.

(2.) Since the introduction of the new rifled arms, exposing cavalry masses to a deadly fire at far greater distances than ever before known, a fire often reaching to the reserves, it seemed doubtful whether the manoeuvring and charging in heavy, compact masses, which formerly rendered cavalry of the line so formidable, would any longer be practicable.

(3.) The comparative cost of this kind of force is so great, that, if it had been raised and kept up on the scale required, the expense of this war, enormous as it has been, would have been vastly augmented. Three years are required for the thorough training and instruction of the men and horses; so that it would not have been until the fourth year of the war that we could begin, even, to reap the fruits of so enormous an outlay.

5. But to carry on any war successfully, what is needed, and is, in fact, indispensable, is an ample force of light cavalry, of a kind requiring comparatively but little time and training to fit it for the various and important duties devolving upon it in the field, and therefore far less expensive than cavalry of the line; and having all the discipline of this latter kind of force, though wanting its perfection of manoeuvre. Every army, or considerable detachment, must have enough of this kind of force with it to furnish what is requisite for Outpost duty, Patrols offensive and defensive, Escorts to trains, Foraging parties, Reconnoissances, and the various other purposes necessarily incidental to operations in the field; and in marches, all Advanced, Rear, and Flank guards should consist, in part, at least, of cavalry. Finally, this description of force is needed for the performance of those arduous, but most valuable, services often rendered by the quasi-independent bodies called Partisan Corps; services usually requiring great celerity of movement.

6. This kind of force being "the eyes and ears of an army," it often contributes powerfully to the success of strategic operations.

In the campaign of 1813, Napoleon complained that, for want of light cavalry, he could get no intelligence of the enemy's movements.

So, in the rebel campaign of 1863, culminating at Gettysburg, General Lee attributed his ignorance of our position and movements, which led to the failure of his operations, to his being destitute of this arm; Stuart's cavalry, on which he depended for information, having got too far away from him.

In Pope's campaign in 1862, the rebels, by their cavalry raid on Catlett's Station, obtained possession of the commanding general's correspondence, plans, and orders from Washington.

On the other hand, whilst keeping us informed of the enemy's movements, an abundant light cavalry, active and well commanded, may be so used as to constitute an impenetrable screen of our own movements from the enemy, as effectual as would be a lofty and impassable mountain range.

Again, if we are greatly inferior to the enemy in cavalry, our own cavalry will have to keep itself within our infantry lines; and the consequence will be that the enemy will obtain control of the entire country around us, and so deprive us of all the supplies it contains.

As, besides this, cavalry is absolutely necessary for the protection of convoys, and, from its celerity of movement, is the kind of force best fitted for guarding our communications, it is evident that the subsistence of an army is dependent, to a great extent, upon this arm.

From what has been said in relation to the three arms, it is evident--

1. That ARTILLERY, within the range of its fire, is powerful in preventing the

enemy's approach to it; but, only to a limited extent, can pursue and drive the enemy from his position; and that its function is, therefore, mainly DEFENSIVE.

2. That CAVALRY, by the impetuosity of its charge, is peculiarly fit for driving the enemy from his position; but, remaining in position itself, has but feeble power to prevent the enemy's approaching it; and this, only by its carbine and pistol fire, which is far from effective; and that its function is, therefore, mainly OFFENSIVE.

3. That INFANTRY has great power, both in keeping the enemy at a distance by its fire and in driving him from his position with the bayonet; and that this arm is, therefore, both OFFENSIVE and DEFENSIVE.

4. That although artillery is mainly a defensive arm, it plays an important offensive part in the powerful assistance it renders to infantry, in shattering and disorganizing the enemy's masses; thereby opening the way for our attacking columns.

5. That although cavalry is mainly an offensive arm, its defensive value is also very great in the protection it affords, in various emergencies, to the other arms, by its actual charge, or by its threatening position.

The special parts usually played in battle by the three arms respectively, may be briefly stated thus:--

Artillery prepares the victory; Infantry achieves it; Cavalry completes it, and secures its fruits.

THE END.